RCA **CRD** RESEARCH PUBLICATIONS
GENERAL EDITOR GILLIAN CRAMPTON SMITH
EDITOR GILES LANE

FIRST PUBLISHED IN 1999 BY THE
ROYAL COLLEGE OF ART
COMPUTER RELATED DESIGN
RESEARCH STUDIO
KENSINGTON GORE
LONDON SW7 2EU
UNITED KINGDOM

www.crd.rca.ac.uk/research

BOOK DESIGNER PAUL FARRINGTON

COVER PHOTO JULES SELMES
TYPE MATRIX, BONE A, SANSCOUNTER STEW, SANSCOUNTER STIR & INTERSTATE **PAPER** TEXT – JAMES McNAUGTON®, 130GSM
CYCLUS PRINT, END PAPERS – 60 GSM CHALLENGER BANK **COVER** JAMES McNAUGTON®, 380 MICS CHALLENGER PULPBOARD
PRINTED IN THE UK BY GEOFF NEAL LITHO, MIDDLESEX

BRITISH LIBRARY CATALOGUING-IN-PUBLICATION DATA: A CATALOGUE
RECORD FOR THIS BOOK IS AVAILABLE AT THE BRITISH LIBRARY

ISBN: 1 874175 27 6

THIS PUBLICATION HAS BEEN GENEROUSLY SUPPORTED BY
INTERVAL RESEARCH CORPORATION

Royal College of Art
Postgraduate Art & Design

HERTZIAN TALES
ELECTRONIC PRODUCTS, AESTHETIC EXPERIENCE AND CRITICAL DESIGN

ANTHONY DUNNE

RCA **CRD** RESEARCH

FOREWORD

Initially funded by a generous grant from the Silicon Valley company, Interval Research Corporation, the Computer Related Design Research Studio at London's Royal College of Art was founded to investigate how the skills and knowledge of artists and designers might be applied to the design of information technology systems and products. Bringing together a variety of perspectives — architecture, industrial and graphic design, psychology, engineering and fine art — the Studio now collaborates on projects with companies and organisations worldwide. This book is the first of several bringing the work of the Studio to a wider audience.

In the 1950s Reyner Banham, in THEORY AND DESIGN IN THE FIRST MACHINE AGE, was already describing the arrival of the Second, "the age of domestic electronics and synthetic chemistry". If the First Machine Age was heroic (cars, planes, heavy industry, speed), in the Second, the age of small machines (vacuum cleaners, washing machines, electric mixers) made possible by electricity and advances in control technology, the machine was literally domesticated, an ingredient of everyday experience. Forty years later we live in the Third Machine Age, an age of electronic devices that hardly seem like machines at all. Made possible first by the transistor and later by the silicon chip, computers were initially designed by technologists as technical tools for other technologists. They then entered the office. Only recently has computer technology, converging with telecommunications technology, invaded the home.

The work of Dr Anthony Dunne, who trained and has practiced as an industrial designer, is innovative, haunting, but always benevolent and clear-sighted. In HERTZIAN TALES he reflects on the historical and conceptual context of this work in the light of the electronic infiltration of everyday life.

This vast potential of 'electronic omnipresence' has hardly begun to be recognised. Manufacturers tell us our toaster is "electronically controlled", but it still just toasts bread. Our washing machine may be 'electronic' but, though with less human effort, it does no more than the washboard and wringer. Yet electronic machines could serve altogether new purposes. If the first two Machine Ages were truly mechanical, extensions of our musculature, electronic machines can extend our minds, our subjective relationship with the world, our active relationship with other people.

Back in the Second Machine Age, in 1951, my family bought its first fridge — a gas one, because it was cheaper to run. Our previous means of delaying food decay had been earthenware coolers soaked in water, marble slabs, and leaving food outside in winter. Our new fridge kept things cold for longer. It also signified for us, however, the end of post-war austerity and the family's progressive optimism. Objects of use, in short, seldom have solely practical significance, but also carry ritual and symbolic meaning. This has long been understood by manufacturers of consumer products (our fridge was vaguely streamlined to withstand the hurricane pace of our new-found modernity). But in the design of computer technology this aspect has until recently been either considered simplistically, as a matter of 'executive' styling for instance, or ignored.

Objects change not just how we see ourselves but, moreover, how we live life. The domestic fridge reduced the need for food to be bought daily and cooked twice-daily. Partly in consequence, female employment patterns changed and family life began to lose its ritual ceremony of solidarity, hierarchy and manners: the shared mealtime.

To suggest how electronic objects might enrich rather than impoverish the lives we now share with them, Dunne proposes an "aesthetics of use": an aesthetics which, through the interactivity made possible by computing, seeks a developing and more nuanced cooperation with the object — a cooperation which, it is hoped, might enhance social contact and everyday experience. Such an aesthetics, clearly, attends less to how an object looks, the traditional concern of product aesthetics, than to how it behaves.

Unsurprisingly, the development of new electronic products is currently led by technology companies and what they think will sell. To reduce risk, this development tends to be incremental: customers will probably like a product similar to a predecessor that has already sold well, and its manufacture should present few new problems. But this conservatism, though understandable, tends only to offer ever more intelligent toasters. Architecture has a long tradition of the 'ideas competition' whose entries are not usually intended to be built but to publicly disseminate radical ideas about how architecture, and possibly the life it accommodates, might be differently conceived. The world of electronic product design needs a similar speculative arena, to imagine possible and impossible futures with computer technology, and to extend what people might find enjoyable to live

with and what manufacturers could imagine making and selling. The Computer Related Design Research Studio aims to be part of this arena. Dunne + Raby, founder members of the Studio, contribute to it a distinctive and powerful voice.

HERTZIAN TALES explores a space between fine art and design, showing how designers can use fine-art means — provoking, making ambiguous, making strange — to question how we cohabit with electronic technology and to probe its aesthetic potential. Dunne describes previous approaches to the design of electronic objects, and, more generally, how other disciplines — semiology, anthropology, design and literature — have conceived the cultural role of the object. He then considers the "aesthetics of use" and what he calls "psychosocial narratives" and "real fictions", concentrating less on sculpture, the traditional reference for industrial designers, and more on the "complicated pleasures" of literature and film.

Dunne believes that "the most difficult challenges for designers of electronic objects now lie not in technical and semiotic functionality, where optimal levels of performance are already attainable, but in the realms of metaphysics, poetry and aesthetics, where little research has been carried out." In these realms he is a bold pioneer.

Gillian Crampton Smith
Professor of Computer Related Design
Royal College of Art

CONTENTS

ACKNOWLEDGMENTS

I would like to thank the following for their help throughout this research project: Kei'ichi Irie, John Thackara and Daniel Weil for their encouragement and advice while getting the project underway; Gillian Crampton Smith for her supervision and encouragement during the development of the PhD thesis; Phil Tabor for his invaluable critical observations and advice during the final drafts of the thesis; Alex Seago for his helpful comments on early drafts; CRD staff and students for thorough discussion and debate during the development of this project; the many people who provided opportunities to test the ideas in this book through lectures, workshops and teaching; Spencer Childs for technical advice about the world of radio; Jayne Roderick for sound design; Dan Sellars for the PILLOW TALK video; Lubna Hammoud for photographic portraits of the THIEF OF AFFECTIONS; Shona Kitchen for detective work and picture research; and Fiona Raby my partner and collaborator for inspiration, reality checks, and endless support.

This project has been generously supported by Interval Research Corporation.

Whereas architecture and furniture design have successfully operated in the realm of cultural speculation for some time, product design's strong ties to the marketplace have left little room for speculation on the cultural function of electronic products. As ever more of our everyday social and cultural experiences are mediated by electronic products, designers need to develop ways of exploring how this electronic mediation might enrich peoples' everyday lives.

HERTZIAN TALES explores the way critical responses to the ideological nature of design(01) can inform the development of aesthetic possibilities for electronic products. It focuses on the role they play in shaping our experience of inhabiting the 'electrosphere', looking beyond the quality of our relationship with objects themselves to the aesthetics of the social, psychological and cultural experiences they mediate.

The primary purpose of this book is to set the scene for relocating the electronic product beyond a culture of relentless innovation for its own sake, based simply on what is technologically possible and semiologically consumable, to a broader context of critical thinking about its aesthetic role in everyday life.

The project proposes an approach that uses the design of conceptual electronic products as a way of provoking complex and meaningful reflection on the ubiquitous, dematerialising and intelligent artificial environment we inhabit.

The origins of HERTZIAN TALES as a project lie in two areas: firstly a frustration with the limited role currently played by industrial designers (compared to those of engineers and marketeers) in the development of new electronic products; and secondly a belief that design, too, has much to contribute as a form of social commentary, stimulating discussion and debate amongst designers, industry and the public about the quality of our electronically mediated life.

The book aims to map out the current technologically-informed environment of ideas about electronic objects and to understand industrial design's potential within it – developing a position that is both critical and optimistic.

Industrial design is not art, but neither is it purely a business tool. While mainstream industrial design is comfortable using its powerful visualisation capabilities to propagandise desires and needs designed by others, thereby maintaining a society of passive consumers, design research in the aesthetic and cultural realm should draw attention to the ways products limit our experiences and expose to criticism and discussion their hidden social and psychological mechanisms.

I believe strongly in the potential of industrial design as applied art, or industrial art, to improve the quality of our relationship to the artificial environment, and in industrial design's potential, at the heart of consumer culture, to be subverted for more socially beneficial ends. In order to achieve this, research is needed into an expanded notion of design aesthetics that includes more poetic and metaphysical relationships with the artificial environment of technological artefacts.

12 PREFACE

Although much has been written about the inter-relationship of technology and culture, few
sources address design specifically — and there is a surprising lack of examples of such
research within industrial design. Even the cultural and aesthetic experiments of design
groups like Memphis, or more recently Droog Design, rarely touch on electronics.
(Although arguably the general changes brought about through electronic technologies
have played a part in motivating such work). Because of this, this book draws many
examples from other fields such as fine art and architecture.

Not surprisingly, research such as this is difficult to situate in a purely commercial context.
Academia provides a space for developing ideas and approaches — 'hypotheses for action'
— but, as in technical research, an industrial context better suits the complexities of
realisation and implementation. An academic setting can also provide a new context for
design outside commercial practice, distinct from the more established critical
approaches of architecture and fine art.

Andrea Branzi and other experimental designers of the 1960s and 1970s addressed the role played
by design in poetic modes of inhabitation and, guided by an architectural perspective,
developed provocative research approaches and positions focusing on the expressive and
linguistic possibilities of new materials and surfaces. (02) During the early 1980s Daniel
Weil produced a group of design proposals for radios applying some of the concerns of
Italian radical design to electronic products. More recently, Ezio Manzini outlined a role
for the designer that offers a fresh perspective building on earlier Italian design thinking.
He suggests that the days of the design visionary are over and a weariness with utopian
visions has set in. Instead, he suggests that the designer use his or her skills to visualise
alternative future scenarios in ways that can be presented to the public, thus enabling
democratic choices between the futures people actually want. Designers could then set
about achieving these futures by developing new design strategies to direct
industry to work with society.

But in the HERTZIAN TALES project, the designer, like J.G. Ballard's writer, (03) no longer knows
anything for certain; all he or she can offer are the contents of his or her own head,
where internal imagination meets the external world of reality. Design is used as a
strategy for linking these two worlds. Its outcome consists of conceptual design proposals

offering a critique of the present through the material embodiment of functions derived from alternative value systems. These 'material tales' are not utopian visions or blueprints — clear-cut modelling of the future is too didactic. Instead they mix criticism with optimism to provide the "complicated pleasure" ⟨04⟩ found in other imaginative media such as film and literature, particularly those that explore boundaries between the real and the unreal.

HERTZIAN TALES consists of two main parts: six essays exploring design approaches for developing the aesthetic and critical possibilities of electronic products outside a commercial context, and five conceptual design proposals expressed as objects, videos and images — by-products of an investigation into a synthesis between practice and theory, where neither practice nor theory leads.

The first chapter, **THE ELECTRONIC AS POST-OPTIMAL OBJECT**, discusses existing design approaches to the development of aesthetic possibilities for electronic objects. Current design approaches aim to optimise the experience of using an object, with the effect of constraining our experience to the prosaic. However, when practicality and functionality can be taken for granted, the aesthetics of the 'post-optimal' object provide a much richer field of investigation.

If user-friendliness characterises the relationship between people and the optimal electronic object, then user-unfriendliness, a form of gentle provocation, could characterise the post-optimal object. **(IN)HUMAN FACTORS** discusses a role for different degrees of user-unfriendliness such as estrangement and alienation in the design of electronic objects.

Design's emphasis on forms of expression and languages of representation, rather than on experience, holds back the potential of electronic objects to provide new levels of aesthetic experience. **PARA-FUNCTIONALITY** investigates the design of function (rather than form) to provide new types of aesthetic experience.

PSYCHOSOCIAL NARRATIVES discusses behaviour as a narrative experience determined by objects, and how the embodiment of unusual psychological needs and desires in electronic objects can encourage the user to experience new narrative situations as a protagonist. These new possibilities complement normality by referring to the world of object misuse and abuse.

REAL-FICTION discusses systems of presentation and consumption for ideas which, unlikely to be mass-produced or even prototyped, exploit the conceptual status of objects as ideas.

HERTZIAN SPACE explores how artists and designers have made links between the invisible environment of electromagnetic radiation and the material culture of objects, and draws attention to the neglected aesthetic dimensions of electromagnetic fields.

HERTZIAN TALES AND SUBLIME GADGETS is a commentary on the design work for this project. It consists of five conceptual design proposals for the electronic as a post-optimal object: **Electroclimates** (with Fiona Raby), **When Objects Dream...**, **Thief of Affections**, **Tuneable Cities** (with Fiona Raby), and **Faraday Chair**. They are not necessarily illustrations of the ideas discussed in earlier chapters, nor are the earlier chapters an explanation of these proposals. They evolved simultaneously and are part of the same design process.

NOTES

(01)
The mainstream view of industrial design serving the narrow commercial interests of industry as opposed to a more general social role for design: developing tools for living.

(02)
The adventurous spirit and humanist vision of Italian design research from the 1960s, 1970s and early 1980s has been an inspiration throughout this project. The design approach developed for electronic objects in this book is indebted to their research.

(03)
"I feel myself that the writer's role, his authority and licence to act, have changed radically. I feel that, in a sense the writer knows nothing any longer. He has no moral stance. He offers the reader the contents of his own head, he offers a set of options and imaginative alternatives. His role is that of the scientist, whether on safari or in his laboratory, faced with a completely unknown terrain or subject. All he can do is to devise various hypotheses and test them against the facts." J.G. Ballard, CRASH, p. 9.

(04)
In his introduction to EINSTEIN'S MONSTERS Martin Amis writes that the purpose of the stories is "no purpose at all — except, I suppose, to give pleasure, various kinds of complicated pleasure".

As new technical developments alter the object and make it 'intelligent', they also set the object on a plane with no prior cultural references... although the physical aspects of these objects are still within the world of materials, their operation and their very state of being is well beyond the manipulation of matter and has more to do with information exchange than with form.

E. MANZINI, THE MATERIAL OF INVENTION

Most designers of electronic objects have responded to this challenge by accepting a role as a semiotician, a companion of packaging designers and marketeers, creating semiotic skins for incomprehensible technologies.

From Banham writing about portable radios in the 1970s, through the plethora of essays on 'product semantics' in the 1980s, to Norman Bolz's 1994 essay THE MEANING OF SURFACE, the treatment of the electronic object as a package for technology, designed to communicate use, cultural meaning, and corporate identity through its surface, has been thoroughly explored. The electronic object accordingly occupies a strange place in the world of material culture, closer to washing powder and cough mixture than furniture and architecture, and is subject to the same linguistic discipline as all package design, that of the sign. It is lost somewhere between image and object and its cultural identity is defined in relation to technological functionalism and semiotics.

01
THE ELECTRONIC AS POST-OPTIMAL OBJECT

This chapter considers three perspectives on the electronic object: **THE ELECTRONIC AS LOST OBJECT** briefly discusses theoretical perspectives, **THE ELECTRONIC AS OBJECT** focuses on design approaches and **THE ELECTRONIC AS POST- OPTIMAL OBJECT** introduces the idea of the 'post-optimal' object. (01)

THE ELECTRONIC AS LOST OBJECT

A TECHNOLOGICAL PERSPECTIVE

From a technological perspective the theories of Jean Baudrillard and Paul Virilio are a stimulating
source of ideas about the effects of electronic technology on the way we experience and
think about ourselves, objects and environments. Their provocative fusions of analysis
and imagery offer a rich inspiration while remaining grounded in reality. But there is a
danger that if designers are seduced by this, their designs will become mere illustrations
of descriptions of electronic objects. Designers of electronic objects are already familiar
with the kinds of technologies analysed by these writers. It is more important
to extend the range of cultural values, building on what is already understood, rather
than illustrating it.

Some writers on the social history of technology present the ideological dimension of everyday
technologies, even if these are often pre-electronic. This is useful to critique the Human
Factors 'community', who have developed a view of the electronic object, derived from
computer science and cognitive psychology, that is extremely influential in the computer
industry, for example Don Norman's THE PSYCHOLOGY OF EVERYDAY THINGS.

A serious problem with the Human Factors approach though, in relation to this project, is its
uncritical acceptance of what has been called by Bernard Waites the "American Ideology",
the ideological legitimation of technology:

*All problems whether of nature, human nature, or culture, are seen as 'technical' problems
capable of rational solution through the accumulation of objective knowledge, in the form
of neutral or value-free observations and correlations, and the application of that
knowledge in procedures arrived at by trial and error, the value of which is to be judged by
how well they fulfil their appointed ends. These ends are ultimately linked with the
maximisation of society's productivity and the most economic use of its resources, so that
technology, in the American Ideology, becomes 'instrumental rationality' incarnate, the
tools of technocracy.*

B. WAITES, EVERYDAY LIFE AND THE DYNAMICS OF TECHNOLOGICAL CHANGE

The result, as the computer industry merges with other industries, is that the optimisation of the psychological fit between people and electronic technology, for which the industry strives, is spreading beyond the work environment to areas such as the home which have so far acted as a counterpoint to the harsh functionality of the workplace. When used in the home to mediate social relations, the conceptual models of efficient communication embodied in office equipment leave little room for the nuances and quirks on which communication outside the workplace relies so heavily.

Writing on electronic art might seem a good source of ideas on the electronic object, but, surprisingly, electronic art has become so technology-driven that it seems concerned only with the aesthetic expression of technology for its own sake. Rather than relating the impact of technology to everyday life, art criticism in this area glamorises technology as a source of aesthetic effect to be experienced in galleries. The exceptions tend to be based on electronic systems rather than objects (for example in the work of Roy Ascott).

A SEMIOTIC PERSPECTIVE

A semiotic approach has been taken by design writers, both at the linguistic level, looking at the way objects can be 'written' and 'read' as visual signs, and at the more general level of the study of consumerism, where semiological analysis of objects as commodities has revealed their part in maintaining what Roland Barthes has called "mythologies". An impressive semiotic analysis of the object is Baudrillard's, FOR A CRITIQUE OF THE POLITICAL ECONOMY OF THE SIGN, which shifts the emphasis on the analysis of commodities away from the production of objects to the consumption of signs.(02) But, as Daniel Miller writes:

While the rise of semiotics in the 1960s was advantages [sic] in that it provided for the extension of linguistic research into other domains, any of which could be treated as a semiotic system, this extension took place at the expense of subordinating the object qualities of things to their word-like properties.

D. MILLER, MATERIAL CULTURE AND MASS CONSUMPTION

A MATERIAL CULTURE PERSPECTIVE

Although there is very little available on the electronic object, the study of material culture is still
of interest because it situates the object firmly within everyday life. Academically, it is
somewhere between anthropology, sociology and ethnology.

Miller claims there is a an "extraordinary lack of academic discussion pertaining to artefacts as
objects, despite their pervasive presence as the context for modern life", and provides an
alternative to the semiological analysis of mass consumption by distinguishing material
culture from language and the study of meaning in order to focus on the physical nature
of artefacts.

In contrast to analyses of the object in relation to consumerism, Mihaly Csikszentmihalyi and
Eugene Roshberg-Halton's, THE MEANING OF THINGS, analyses the meaning of objects in
domestic settings, emphasising their symbolic role. And in THE METAFUNCTIONAL AND
DYSFUNCTIONAL SYSTEM: GADGETS AND ROBOTS, Baudrillard writes about the electronic gadget
as the subject of a science of imaginary technical solutions. Although originally written
nearly thirty years earlier, Baudrillard's analysis of electronic gadgets is far more
stimulating than a more recent analysis, CONSUMING TECHNOLOGIES, by Roger Silverstone
and Eric Hirsch, which is more concerned with descriptive models than Baudrillard's
challenges to the imagination. HERTZIAN TALES is more concerned with 'critical'
theories,(03) and thus in assessing the development of objects not against whether they fit
into how things are now, but the desirability of the changes they encourage.

The value of material culture for this study is that it draws attention to the complex nature of our
relationship to ordinary objects and provides standards against which new electronic
objects can be compared.

A DESIGN PERSPECTIVE

Since the early 1960s a very narrow form of semiotic analysis has dominated design
thinking about the electronic object. Of books written about design from a theoretical
point of view, only John Thackara's, DESIGN AFTER MODERNISM, contains new
perspectives on the electronic object.

FIG. 1.1
Daniel Weil's RADIO IN A BAG
(1983) takes the idea of the designer's
role as a packager of technology
to the extreme.

Books and articles by designers, based on particular projects prove more interesting. Manzini and Susani present a collection of design projects that explore a place for solidity within the fluid world created by electronic technology:

In the fluid world the permanent features we need are no longer there as a matter of course, but are the result of our desire; the 'solid side' in a fluid world, if and when it exists, will be the result of a design.

E. MANZINI & M. SUSANI, THE SOLID SIDE

Their strong emphasis on aesthetics and ecological concerns is a powerful example of design research carried out by practising designers within an intellectual context. Susani has developed a design perspective that locates the electronic object within material culture rather than semiology or electronic media. He writes:

We are lacking a discipline, perhaps an 'objectology', or an 'object ethology', which allows us to analyse and systematise objects and to formulate the rules and codes of their behaviour... a discipline which recovers and updates the interrupted discourse of material culture, in crisis since the world of objects was taken over by the world of products and the world of consumption.

M. SUSANI, THE FOURTH KINGDOM

He also recommends a sensual approach to introducing technology into the home, building on what is already there, and exploring the overlap between the material and immaterial world from an aesthetic and anthropological point of view. He suggests that material culture could offer useful insights to this problem.

A LITERARY PERSPECTIVE

However, the most fruitful reflection on material culture is to be found, not in anthropology or sociology, but in literature concerned with the poetry of everyday objects. In THE POETICS OF SPACE, Gaston Bachelard offers an analysis, influenced by psychoanalysis, that emphasises the poetic dimension of humble furniture such as wardrobes and chests of drawers; Jun'ichiro Tanizaki's IN PRAISE OF SHADOWS considers the Japanese object in relation to shadows and darkness, and the effects of electricity on their appreciation; and Nicholson Baker's novels (such as THE MEZZANINE and ROOM TEMPERATURE) give everyday industrial products significant roles.

The view of objects suggested by literary writers reveals a poetry of material culture that offers a fresh alternative to the formal aesthetic criticism of the art object and to the academic analysis of their meaning as signs. Their objects are firmly grounded in everyday life.

The best writing in this area blends anthropology, sociology and semiology to explore the irrational dimensions of the material culture of everyday life. As the electronic object rarely features in this literature, the discussion in the rest of this book is based mainly on design proposals.

THE ELECTRONIC AS OBJECT

This section discusses four design approaches to the electronic object: packages, fusions, dematerialisation, and juxtaposition. They differ in how each addresses the conflict between the solidity of the object and the fluidity of electronic media. Design is viewed here as a strategy for linking the immaterial and the material.

PACKAGES

Commercial design's approach to the electronic object has been to treat it as a package for electronic technology. An example of this, where the aesthetic and conceptual possibilities of the package are thoroughly exploited, is Daniel Weil's RADIO IN A BAG (1983) **(FIG. 1.1)** which takes the idea of the designer's role as a packager of technology to the extreme. On one level the electronics provide decoration, while on another, their exposure signals a nonchalance towards technology. The radio's literal flexibility expresses the flexible structural relationship between electronic components, and its transparency attempts to demystify the electronic object. It shows that by taking a playful approach to package design and liberating it from product semantics, even the packaging of electronics can yield interesting results. Ironically, part of the critical success of this design, despite being a package is its treatment as a *thing* rather than an image.

FIG. 1.4
Daniel Weil's **CLOCK** (1983),
based on the largest circuit
boards available in the early
1980s is a reaction against
miniaturisation.

FIG. 1.2-3
The first transistor and the **DE FOREST VALVE**: test-rigs
for key electronic components created by inventors working at
the level of both electrons and matter.

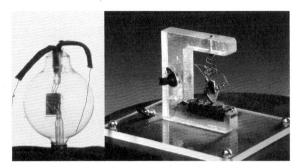

FUSION

> *The logic of computers is expressed in forces that are averages of the behaviour of many*
> *electrons. No machine has ever been so far removed from the world of human experience:*
> *the largest aircraft carriers are still infinitely closer to the human scale than the simplest,*
> *slowest microcomputers.*

D. J. BOLTER, TURING'S MAN

The electronic object is a confusion of conceptual models, symbolic logic, algorithms, software, electrons and matter. The gap between the scales of electrons and objects is most difficult to grasp.

The architect Neil Denari has spoken of the need for the "overcoming of the symbolic", and his view is that architecture must make a connection between the worlds of electromagnetism and spatial inhabitation. But there is greater chance of bridging the gap between electromagnetism and inhabitable space if one where to explore this route through the design of objects rather than buildings.

The first transistor **(FIG. 1.2)** and the DE FOREST VALVE **(FIG. 1.3)** are test-rigs for key electronic components created by inventors who work at the level of both electrons and matter. They organise matter as interacting volumes of electrons, [04] and offer a possibility for reconciling the scales that separate the worlds of electrons and space. But once these prototype elements have been subjected to the extreme rationalisation required by mass-production they become reduced to abstract ultra-miniaturised electronic components. Their modernist poetry, based on truth to materials, is lost.

Closing the gap between the scales of electronics and objects by directly manipulating materials as volumes of electrons is a difficult route for designers. This task is essentially limited to scientists, and even their test-rigs will eventually become miniaturised components. CLOCK by Daniel Weil **(FIG. 1.4)** captures some of this quality — partly a reaction against miniaturisation, its size is based on the largest circuit boards available in the early 1980s. The circuit is composed visually and the wires linking the two main components are made from dining forks. Familiar objects are put into new but natural relationships based on electrical properties.

This approach resembles the way electricity was dealt with in early natural philosophy books which explained electricity in delightfully poetic ways, drawing attention to unusual but real phenomena:

> *The simultaneous development of both kinds of electricity is illustrated by the following experiment: – Two persons stand on stools with glass legs, and one of them strikes the other with a catskin. Both of them are now found to be electrified, the striker positively, and the person struck negatively, and from both of them sparks may be drawn by presenting the knuckle.*

> *J. D. EVERETT, DESCHANEL'S NATURAL PHILOSOPHY PART 3*

The development of 'smart materials' is another area where the gap between the electronic and material is being closed, although primarily for technical reasons. Scientists and engineers are developing new materials, designed at a molecular level, which are responsive, dynamic and almost biological. Although most of these materials are still experimental some, such as electroluminescent laminates(05) and piezoelectric films(06) have been around for several decades.

Manzini explores the implications of designing with these new smart materials:

> *The design of this skin, and therefore of the objects that are made with it, is chiefly the design of interactivity with the environment – a scenario for which we must prepare the stage, the sets, and the actors. Imagining the nature of these 'individual objects' is another new chapter in the history of design.*

> *E. MANZINI, THE MATERIAL OF INVENTION*

Most of Manzini's specially commisioned examples illustrate the miniaturisation arising from integrating previously separate mechanisms and their novel decorative possibilities. However, they do not demonstrate the radical aesthetic potential of these materials to open new channels of communication with the environment of electronic objects.

Only Alberto Meda and Denis Santachiara's STROKE LAMP of 1986 hints at the new relationships between people and machines made possible through new reactive materials. It is controlled by stroking the surface, which is made from an insulating plastic with a copper

FIG. 1.5
Andrea Branzi's **L E A F** light (1988) for Memphis is a rare
example of an application of advanced electrochemical
materials for cultural rather than functional innovation.

circuit deposited on it by a photochemical process similar to that used for printed circuits. Although low-tech, it suggests a sensual and playful interaction with everyday objects that might be extended to more complex interactions as more sophisticated materials become available. Andrea Branzi's LEAF electroluminescent light of 1988 for Memphis is another application of advanced electrochemical materials for cultural rather than functional innovation **(FIG. 1.5)**.

But generally, designers have not exploited the aesthetic dimension of new materials with the same energy that engineers have exploited their functional possibilities (to backlight LCD screens in laptop computers reducing their bulk and weight, for example, or to illuminate escape routes in aircraft so they can be seen through smoke).

Most work in this area does not encourage poetic and cultural possibilities to converge with practical and technical ones. The outcome is a stream of unimaginative proposals. For example, AT&T have applied for a patent for a coating of coloured polymer sandwiched between two thin layers of indium tin oxide that changes colour when a low voltage is fed through it; they plan to use it to enable phones to change colour instead of ringing.

Although combinations of matter and information might eventually lead to interactive surfaces, giving rise to new channels of communication between people and an increasingly intelligent artificial environment of objects, most smart materials are still under development, are expensive and use large amounts of energy to operate. The most interesting materials are not available for design experiments and one must either use simulations or work with widely available but less sophisticated materials to create emblems of what might be.

DEMATERIALISATION

The electronic object is an object on the threshold of materiality. Although 'dematerialisation' has become a common expression in relation to electronic technology, it is difficult to define in relation to the tangle of logic, matter and electrons that is the electronic object.

The CPU of an electronic object is, essentially, physically embodied symbolic logic or mathematics. Its 'material' representation is the circuit and the components it connects. Symbolic logic describes the workings of the 'machine' the object becomes when the program runs. The algorithm is the logical idea behind the program, a strategy that allows symbolic logic to be translated into a programming language (such as C++) and run through the machine, controlling the flows of electrons through its circuitry.

FIG. 1.6
The design group Kunstflug's THE ELECTRONIC ROOM: PROGRAMMABLE
APPEARANCES — SURFACES, APPLIANCES, COMFORT for the
DESIGN TODAY exhibition held at the German Museum of Architecture (1988).

Dematerialisation, therefore, means different things depending on what it is defined in relation to: immaterial/material, invisible/visible, energy/matter, software/hardware, virtual/real. But the physical can never be completely dismissed:

Every symphony has its compact disc; every audio experience its loudspeaker; every visual image its camera and video disc. Behind every outward image or symbol lies mechanical support, and if the immateriality of these images and symbols gives rise to a new approach to the relationship between human being and object, the analysis will be one of the individual's connection with the material support underlying the new culture of immateriality.

A. MOLES, DESIGN IMMATERIALITY: WHAT OF IT IN A POST INDUSTRIAL SOCIETY?

One argument, put forward in the 1980s by the design group Kunstflug, is that values and functions can completely shift from hardware to software, from three to two dimensions, and ultimately to "design without an object". It sounds an untenable and an over-simple critique of materialism, but during the mid-1980s it drew attention to their ideas. They argued for a change in the attitude to the consumption of objects, calling on industry to produce solutions not commodities. "Design without an object" could, as part of a cultural movement, offer an alternative to abstinence from consumption while encouraging "the forsaking of things as objects of desire and covetousness".

In the exhibition DESIGN TODAY, held at the German Museum of Architecture in 1988, Kunstflug offered two examples of this approach: design proposals for THE ELECTRONIC ROOM: PROGRAMMABLE APPEARANCES — SURFACES, APPLIANCES, COMFORT **(FIG. 1.6)** and an ELECTRONIC HAND CALCULATOR **(FIG. 1.7)**. While the room seems only to reinforce stereotypical approaches to the impact of electronics on architectural spaces, the ELECTRONIC HAND CALCULATOR became an icon for "design without an object", defining one extreme position in the debate about the impact of electronic technology on objects.

This interest in dematerialising the object for social and political reasons is echoed by the "info-eco" ideas of Manzini, Susani and Thackara who argue that, by focusing on experiences rather than objects, electronic technology can provide services currently offered through discrete products. In the Info-eco Workshops held at the Netherlands Design Institute in 1995, participants developed scenarios on themes such as 'Beyond Being There'. Dematerialisation was used to investigate hypothetical situations in limited scenarios and

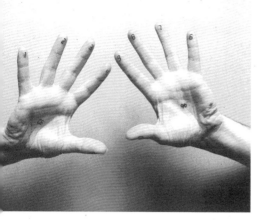

FIG. 1.7
Kunstflug's Electronic **HAND CALCULATOR**
for the **DESIGN TODAY** exhibition held at
the German Museum of Architecture (1988):
an icon for "design without an object".

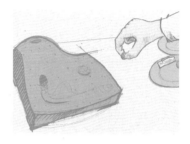

FIG. 1.8
In Durrell Bishop's design for a telephone answering
machine (1992) a small ball is released each time a
message is left – physical representations of the pieces
of information left in the machine.

discover how information technology might satisfy needs normally fulfilled materially. For instance, telematic tools were proposed where the quality of experience they offered would reduce the desire to travel – digital information being easier to move than matter. (Reports detailing the results of the workshops are available at: www.design-inst.nl/).

In the introduction to the 1994 Ars Electronica festival in Linz, Peter Weibel describes a another form of dematerialisation, "intelligent ambience". It arises from shifting emphasis from the 'machine' to its 'intelligence', and distributing that intelligence throughout an environment:

Machine intelligence will serve to make the environment more efficient and more intelligent so that it will be able to respond more dynamically and interactively to human beings. The realisation of the concepts of computer aided design and virtual reality will thus be followed by the realisation of computer aided environment and intelligent, interactive, real surroundings. The latter will be referred to as intelligent ambience – an environment based on machine intelligence. One could say: from Tron house to the Tron ambience.

P. WEIBEL, INTELLIGENT AMBIENCE

Weibel's observations fall between two other views of dematerialisation. The first belongs to the Human Factors world and has been referred to as 'ubiquitous computing' and is the subject of much research. Dematerialisation is seen as a way of providing 'transparent' interfaces for computers by embedding the technology in familiar objects and environments and introducing a high degree of automation. At the other extreme, is Design Primario, [07] where design effort shifts from hardware to software, and controls levels of light, sound and temperature to provide sensual environmental qualities. But the aesthetic possibilities of this form of dematerialisation have been best exploited by architects: Toyo Ito's design for his DREAMS ROOM at the Victoria and Albert Museum in London, was partly motivated by a desire to extend this approach to include information (which he referred to as "active air").

Another form of dematerialisation is defined by electronic objects' role as interfaces. With these objects the interface is everything. The behaviour of video recorders, televisions, telephones, and faxes is more important than their appearance and physical form. Here

FIG. 1.9
Fiona Raby's **TELEMATIC BALCONY**
(1995) is an example of an approach to
electronic objects where no effort is made
to reconcile the different scales of the
electronic and the material.

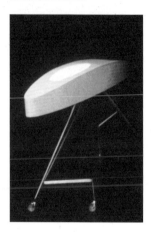

design centres on the dialogue between people and machines. The object is experienced as an interface, a zone of transaction. Although most work in this area tends to reduce the object to a 'graphical user interface', a screen, designers are beginning to explore the full potential of the 'thingness' of the object. The product becomes virtualised and is represented by a set of physical icons and their various permutations. This could lead to more sensual interfaces than screens and offer new aesthetic qualities.

The work of Durrell Bishop offers a vision of what this might mean: existing objects are used as physical icons, material representations of data that refer to both the pragmatic and poetic dimensions of the data being manipulated. The objects and the electronic structure need have nothing in common. For example, in his design for a telephone answering machine **(FIG. 1.8)** small balls are released each time a message is left. These balls are representations of the pieces of information left in the machine, allowing direct interaction between the owner and the many possibilities an answerphone offers for connecting to telephone and computer systems. If the caller leaves a number the ball can automatically dial it; if the message is for somebody else the ball can be placed in his or her personal tray. Although applied very practically, Bishop's thinking engages with the cultural context in which the technology is used. An 'aesthetics of use' emerges.

The material culture of non-electronic objects is a useful measure of what the electronic object must achieve to be worthwhile but it is important to avoid merely superimposing the familiar physical world onto a new electronic situation, delaying the possibility of new culture through a desperate desire to make it comprehensible.

JUXTAPOSITION

How can we discover analogue complexity in digital phenomena without abandoning the rich culture of the physical, or superimposing the known and comfortable onto the new and alien? Whereas dematerialisation sees the electronic integrated into existing objects, bodies and buildings, the *juxtaposition* of material and electronic culture makes no attempt to reconcile the two: it accepts that the relationship is arbitrary, and that each element is developed in relation to its own potential. The physical is as it always has been. The electronic on the other hand is regarded only in terms of its new functional and aesthetic possibilities; its supporting hardware plays no significant part.

Fiona Raby's telematic BALCONY **(FIG. 1.9)** demonstrates how the contradictory natures of electronic and material cultures can co-exist. The balcony provides access to an open telephone line linking two or three places. Its physical form provides a focal point and support for leaning on, while an ultrasonic sensor detects the approach of users and slowly clears the line. There is no point trying to integrate the physical support and the ultrasonic field, to

collapse one into the other, forcing the physical to represent the electronic or to disappear completely so that only electronic effects remain. Juxtaposition allows the best qualities of both to co-exist, each with its own aesthetic and functional potential. Technology can be mass-produced whereas the object can be batch-produced. No effort need be made to reconcile the different scales of the electronic and the material. They can simply coexist in one object. They can grow obsolete at different rates as well. Robert Rauschenberg's ORACLE of 1965 **(FIG. 1.10)** has had its technology updated three times over thirty years but its materiality and cultural meaning remain unchanged. Cultural obsolescence need not occur at the same rate as technological obsolescence.

Perhaps the 'object' can locate the electronic in the social and cultural context of everyday life. It could link the richness of material culture with the new functional and expressive qualities of electronic technology.

In Philips' 1996 VISION OF THE FUTURE project, a more subtle awareness of the value of material culture has entered the mainstream of design thinking and may well soon enter the marketplace and everyday life. The project consists of over 100 design proposals for products for five to ten years in the future. But this awareness is primarily expressed in this project by references to existing object typologies for example, hi-tech medical kits in the form of medicine cabinets, rather than by radically new hybrids. The designers focus more on practical needs, the electronic qualities are not fully exploited, and the types of objects proposed are already familiar from student degree shows. But the designs do achieve a new visual language, sensual, warm and friendly. They are well-mannered and socially competent. In these projects the electronic object has reached an optimal level of semiotic and functional performance.

THE ELECTRONIC AS POST-OPTIMAL OBJECT

The most difficult challenges for designers of electronic objects now lie not in technical and semiotic functionality, where optimal levels of performance are already attainable, but in the realms of metaphysics, poetry and aesthetics where little research has been carried out:

This is what differentiates the 1980s from 1890, 1909, and even 1949 – the ability of industrial design and manufacturers to deliver goods that cannot be bettered, however much money you possess. The rich find their exclusivity continuously under threat...

Beyond a certain, relatively low price (low compared with other times in history) the rich cannot buy a better camera, home computer, tea kettle, television or video recorder than you or I. What they can do, and what sophisticated retailers do, is add unnecessary 'stuff' to the object. You can have your camera gold plated.

P. DORMER, THE MEANINGS OF MODERN DESIGN

The position of this book is that design research should explore a new role for the electronic object, one that facilitates more poetic modes of habitation: a form of social research to integrate aesthetic experience with everyday life through 'conceptual products'.
In a world where practicality and functionality can be taken for granted, the aesthetics of the post-optimal object could provide new experiences of everyday life, new poetic dimensions.

NOTES

(01)
The idea of a 'post-optimal' object arose from a general observation made by Marco Susani during a workshop at the Netherlands Design Institute in 1996. His suggestion, that products in general had reached an optimal level, and that designers should turn there attention towards cities and urban environments which were far from optimal, suggested to me that there must be something beyond the optimal. The idea of a post-optimal object could rescue design objects concerned with ideas, from the clutches of the art world, while maintaining a relationship to design

(02)
For more on the analysis of objects as commodities see A. Appadurai, "Introduction: Commodities and the Politics of Value", in THE SOCIAL LIFE OF THINGS: COMMODITIES IN CULTURAL PERSPECTIVE.

(03)
"Scientific theories have as their aim or goal successful manipulation of the external world; they have instrumental use. If correct, they enable the agents who have mastered them to cope effectively with the environment and thus pursue their chosen ends successfully. Critical theories aim at emancipation and enlightenment, at making agents aware of hidden coercion, thereby freeing them from that coercion and putting them in a position to determine where their true interests lie." R. Geuss, THE IDEA OF A CRITICAL THEORY, pp. 55-56.

(04)
"Electrons are the smallest of these particles and each one carries the smallest amount of negative electricity. In most materials and especially in good insulators like glass or plastics electrons are held firmly in place by heavier, positively charged protons. Some materials, mainly metals, contain electrons that have enough energy to move about allowing the electrons to transport electricity from one place to another, making metals good conductors. When metals are heated, their electrons are given even more energy, sometimes causing them to completely break free from the metal. If they are freed in a vacuum where there is nothing for them to collide with, they can be guided by electricity or magnetism to form controlled electric current. This is the basic principle of the valve and the cathode ray tube." R. Bridgman, ELECTRONICS, p. 26.

(05)
Thin films of polymer material in which particles of doped zinc sulphide have been absorbed, and which, inserted into a sandwich of other protective materials and subjected to alternating current, emit uniform luminosity across their entire surface.

(06)
These transform mechanical impulses into electric impulses or vice versa and are widely used to produce sensors, actuators, microphones and loudspeakers.

(07)
A group of theoretical and design experiments carried out by Andrea Branzi and Clino Trini Castelli during the mid-1970s. Emphasis was placed on "experiences of space that are not directly assimilable to the constituent qualities of an environment or an object, but are linked instead with the physical perception of space, i.e. with its 'bodily' consumption. In this way new attention was paid to the user's real sensitivity of perception, bound up more closely with the direct consumption of soft structures than with the grasping of an architectural composition and its sophisticated allegories of form". A. Branzi, THE HOT HOUSE, pp. 97-98.

Am I a man or a machine? There is no ambiguity in the traditional relationship between man and machine: the worker is always, in a way, a stranger to the machine he operates, and alienated by it. But at least he retains the precious status of alienated man. The new technologies, with their new machines, new images and interactive screens, do not alienate me. Rather, they form an integrated circuit with me.

J. BAUDRILLARD, XEROX AND INFINITY

n design, the main aim of interactivity has become user-friendliness. Although this ideal s accepted in the workplace as improving productivity and efficiency, its main assumption, that the way to humanise technology is to close the gap between people and machines by designing 'transparent' interfaces, is problematic, particularly as this view of nteractivity has spread to less utilitarian areas of our lives.

Interactive user-friendliness' ... is just a metaphor for the subtle enslavement of the human being to 'intelligent' machines; a programmed symbiosis of man and computer in which assistance and the much trumpeted 'dialogue between man and the machine' scarcely conceal the premises:... the total, unavowed disqualification of the human in favor of the definitive instrumental conditioning of the individual.

P. VIRILIO, THE ART OF THE MOTOR

This enslavement is not, strictly speaking, to machines, nor to the people who build and own them, but to the conceptual models, values, and systems of thought the machines embody. User-friendliness helps to naturalise electronic objects and the values they embody. For example, while using electronic objects the use is constrained by the simple generalised model of a user these objects are designed around: the more time we spend using them the more time we spend as a caricature. We unwittingly adopt roles created by the Human Factors specialists of large corporations. For instance, camcorders have many built-in features that encourage generic usage; a warning light flashes whenever there is a risk of 'spoiling' a picture, as if to remind the user that they are about to become creative and should immediately return to the norm. Raising the distance between people and electronic objects, sensitive scepticism might be encouraged, rather than unthinking assimilation of the values and conceptual models embedded in electronic objects. I am not arguing for a way of designing that is free from ideological content but, rather, one that draws attention to the fact that design is *always* ideological. User-friendliness helps conceal this fact. The values and ideas about life embodied in designed objects are not natural, objective or fixed, but man-made, artificial and muteable.

oter looks at 'poeticising' the distance between people and electronic objects through estrangement' and 'alienation', locating interactivity between transparency and opaqueness, the pet and the alien, prose and poetry. The first section looks at the origins of user-friendliness in Human Factors and how it manifests itself in design approaches; the second, **TRANSPARENCY**, discusses the implications of closing the distance between people and machines; and the third, **(IN)HUMAN FACTORS**, looks at

USER-FRIENDLINESS

Manuel DeLanda situates the origins of the man-machine interface within a military context:

It is at the level of the interface that many of the political questions regarding Artificial Intelligence are posed. For instance, one and the same program may be used to take human beings out of the decision-making loop, or on the contrary, [be] interfaced with them so as to create a synergistic whole. It is the design of the interface which will decide whether the machinic phylum will cross between man and machines, whether humans and computers will enter into a symbiotic relationship, or whether humans will be replaced by machines. Although the centralizing tendencies of the military seem to point to a future when computers will replace humans, the question is by no means settled.

MANUEL DELANDA, WAR IN THE AGE OF INTELLIGENT MACHINES

DeLanda writes that research into interactivity between people and computers began with the military's desire to visualise data held in computers, and that interactivity went much further than it intended, giving people total control over their machines. Although scientists such as Doug Engelbert, Alan Kay, J. C. R Licklider and Murray Turoff managed to gain control of the evolution of computers from the military, developing a vision of interactivity as a partnership between people and machines acted out on the computer screen, they were unable to introduce them into everyday life. It was hackers like Steve Wozniak and Steve Jobs who eventually managed to translate these ideas into a machine that could compete in the marketplace against large corporations like IBM and establish a new model of interactivity.

While interactivity made huge leaps forward before its entry into everyday life through the marketplace, once the computer became a successful mass-produced object, innovation in interactivity shifted from hardware to software, and evolved around screens, keyboards and mouse-like input devices.

THE HUMAN FACTORS APPROACH

These days most work on the development of interfaces is by engineers and scientists working for large corporations and universities, and comprising the Human Factors community.

FIG. 2.1
The approach taken by Ettore Sottsass for the design of the
ELEA 9003 computer for Olivetti (1959) is very different from
the 'user-friendly' approach taken by the Human Factors community
which reduces the relationship between people and technology to
a level of cognitive clarity.

Although mainly concerned with computers, other electronic objects are becoming subject to this approach, particularly as designers have, so far, been unable to develop convincing alternatives.

In a review of THINGS THAT MAKE US SMART by cognitive psychologist Don Norman, Rick Robinson offers remarks about Norman's view of design that are applicable to the Human Factors community in general. Robinson argues that Norman's approach results in products that will not confuse or disappoint, (which is clearly not enough). His main criticism is that it:

misses the essential connection between the power of objects to affect the way in which the world is seen and the mechanism through which that happens. Paradoxically, user-centredness is not just figuring out how people map things, it absolutely requires recognising that the artefacts people interact with have enormous impact on how we think. Affordances, to use Norman's term, are individually, socially, and culturally dynamic. But the artefacts 'do not merely occupy a slot in that process, they fundamentally shape the dynamic itself.

R. ROBINSON, IN D E S I G N I S S U E S *10 (i)*

DESIGN/AESTHETIC MANIFESTATIONS

In the Human Factors world, objects, it seems, must be understood rather than interpreted. This raises the question: are conventional notions of user-friendliness compatible with aesthetic experience? Perhaps with aesthetics, a different path must be taken: an aesthetic approach might subsume and subvert the idea of user-friendliness and provide an alternative model of interactivity.

The reduction of the relationship between people and technology to a level of cognitive clarity by the Human Factors community contrasts with the approach taken by Ettore Sottsass in the late 1950s for the design of the ELEA 9003 computer for Olivetti **(FIG. 2.1)**:

It was immediately obvious in the first years in which I worked on the ELEA that in the design of certain gigantic instruments, as electronic machines were then, or in the design of groups of machines which have a logical and operational relationship between each other, one ends up immediately designing the working environment; that is, one ends up conditioning the man who is working, not only his direct physical relationship with the instrument, but also his very much larger and more penetrating relationship with the whole act of work and the complex

FIG. 2.2
Marco Zanuso & Richard
Sapper's Television for Brion
Vega was a sophisticated
expression of a new role for the
skin of an object, with very
different characteristics in both
its states. Switching it on or off
transformed it from familiar to
mysterious object.

FIG. 2.3
Matthew Archer's miniature computer is one
of many projects produced in the Industrial
Design department at the Royal College of Art
during the 1980s which exploited the new
freedom offered to design by the fluid
qualities of electronic technology.

mechanisms of physical culture and psychic actions and reactions with the environment in which he works, the conditionings, the liberty, the destruction, exhaustion and death.

E. SOTTSASS (01)

Although Sottsass' design for a computer clearly derives from a poetic model of people, few designers have developed such powerful aesthetic responses to electronic objects. An exception was Marco Zanuso, whose television for Brion Vega **(FIG. 2.2)** was designed with Richard Sapper in 1969 during the high point of the Italian Radical Design movement, and was at the cutting edge of design thinking, a new expression of an everyday electronic product. It took the notion of the black box to the limit, revealing the magic of technology by dissimulating its functional nature. The whole object became a screen, working equally well aesthetically, on or off. Its minimal black form receded when the television image was shown, and became a pure object when it was switched off. It was concerned with not so much form or even material, but rather the problem of an object with different characteristics in both of its states. It represented a sophisticated expression of a new role for the skin of an object.

Despite this, and because the mechanical design of electronic objects gives few clues to their operation, the problem they posed to most designers soon reduced to one of packaging. But for more experimental designers, the image of the black box became the starting point for exploring new languages of representation rather than interactivity.

REPRESENTATION

During the early 1980s in the department of Industrial Design at the Royal College of Art (RCA) many innovative projects were produced **(FIG. 2.3)** which exploited the new freedom offered by the fluid qualities of electronic technologies, although most were still concerned more with representation and interpretation than function or interactivity. As a group these works are impressively diverse, original and fresh. Implying no clear manifesto or philosophy, but rather reflecting the individual personalities and interests of the designers. They explore how different languages of form map onto electronic technologies by reinterpreting existing products. Many of the presentation models were

FIG. 2.4
Semiotics and semantics were used by 1980s
designers as a framework for analysing the way
industrial designers could use form to express implicit
meanings: for instance the flow of air in this fan
heater (1981) by Winfried Scheuer.

simplified, intended to communicate ideas about form and representation rather than manufacture and practicality. The most relevant work from this era, by Weil, is discussed later in this chapter.

PRODUCT SEMANTICS

During the 1980s 'product semantics' began to influence thinking about electronic products. Semantics and semiotics were originally used by linguists to understand the structure of language and how it conveys meaning, and later by film theorists (often combined with psychoanalysis – Laura Mulvey, for example) to analyse how codes and conventions work. In design they were used to analyse the way form could be used to express implicit meanings: the flow of air in a fan heater for instance **(FIG. 2.4)**.

Cranbrook Academy's industrial design course developed this approach, led by Michael and Katherine McCoy. From the mid-1980s its students fed the international design press a steady stream of conceptual designs for electronic products. In 1987 one of them, Lisa Krohn (with Tucker Viemeister), won a competition to promote and exploit the versatile properties of plastics with her design for an answerphone **(FIG. 2.5)**. The versatility of plastics in this instance is in the area of linguistic expression:

A combination of telephone and answering machine which transcribes and thermally prints any messages, its modern streamlined appearance uses a book format with the pages serving as switches for the different functions.

FORM FINLANDIA, *PROMOTIONAL LEAFLET FOR DESIGN COMPETITION (1987)*

Such literal use of analogy results in metaphors with a single meaning. Products depict what they do, limiting the viewer's interpretation of the electronic object to the designer's, and, although sometimes the link made between groups of objects is ingenious, the power of these borrowed images to sustain interest is weak – they are the material equivalent of one-liners. Once the viewer grasps the connection there is little else to engage with.(02)

The new forms are just as vigorously tied to their signifieds as the old ones, albeit signifieds extrinsic to the object, based in a cultural frame of reference. To use pre-existing patterns of knowledge to define a new technology's possibilities for conveying meaning is not far

FIG. 2.5

Lisa Krohn's design (with Tucker Viemeister) for an answerphone (1987) shows how a literal use of analogy results in metaphors with a single meaning. Products become depictive of what they do, limiting the viewer's interpretation of the electronic object to the designer's.

removed from the Victorian use of Corinthian columns to support beam engines, design holds back the potential of electronics to provide new aesthetic meanings:

Official culture still strives to force the new media to do what the old media did. But the horseless carriage did not do the work of the horse; it abolished the horse and did what the horse could never do.

M. MCLUHAN, COUNTERBLAST

TRANSPARENCY

Because the mimetic approach has greatly affected mainstream thinking about electronic objects, most designs for interfaces with electronic products draw on familiar images and clichés rather than stretching design language. Nothing is what it appears, but simply an allusion to something we are already familiar with. Designers using existing codes and conventions to make new products more familiar, often unconsciously reproduce aspects of the ideology encoded in their borrowed motifs. The easy communication and transparency striven for by champions of user-friendliness simply make our seduction by machines more comfortable.

BIOMORPHISM

The trend for forms of biomorphic expression, particularly in cameras and other portable devices, can be seen as expressing either an uncritical desire to absorb technologies into the body, a wish to be a cyborg, or, more optimistically, a need to mould technology to the body. But this need for symbiosis does not have to be expressed through the clichéd language of bio-form; after all, the symbiosis yearned for is often mental not physical. An engaging, if conservative, image of this desire for symbiosis between people and the environment of electronic artefacts is provided by the series of kitchen tools designed by Marco Susani and Mario Trimarchi for the 1992 Milan Triennale. A mixture of abstract form and familiar materials, they neither pretend to have always been there nor are they completely alien. **(FIG. 2.6)**

For extreme expressions of this wish for transparency or symbiosis we need to look outside the design field, at the work of the artist Stelarc. He describes a synthetic skin that, absorbing

FIG. 2.6
Marco Susani & Mario Trimarchi's **NEW TOOLS** **(FOR THE KITCHEN)** for the 1992 Milan Triennale demonstrate that the need for symbiosis does not have to be expressed through the clichéd language of bio-form; after all, the symbiosis yearned for is often mental not physical.

oxygen through its pores and efficiently converting light into chemical nutrients, would make our internal organs redundant and allow them to be removed to create room for more useful technological components. In a performance at the Doors of Perception 3 conference in Amsterdam in 1995, remote viewers were able to manipulate his body into positions that represented letters; a computer program allowed sequences to be made up forcing the artist, through electrical stimulation of his muscles, to enact a bizarre semaphore. In an earlier piece, THIRD HAND, he wrote single words with a third artificial hand strapped to one of his own **(FIG. 2.7)**, activated by the EMG signals of the abdominal and leg muscles, while his real arm was remotely controlled and jerked into action by two muscle stimulators. Stelarc's work illustrates one vision of cyborgs. His work explores the interplay between self-control of the body and its control by the technological logic embodied in prosthetic devices.

PETS

If the desire for familiarity is applied to more complex machines with a potential for autonomous behaviour, we could find ourselves living in a bestiary of technological 'pets', or zoomorphic electronic objects. Although there is plenty of potential for new aesthetic experiences through the expression of electronic objects' behaviour, this area is already dominated by an over-simple mimicry of human and animal behaviour. The aesthetic experience they give rise to is based on recognition rather than perception. [03] The users experience something familiar rather than new, so are conditioned to accept things as they are. Rather than being stimulated to modify their ideas about reality, the users become part of a behavioural 'circuit':

The famous Japanese car that talks to you, that 'spontaneously' informs you of its general state and even of your general state, possibly refusing to function if you are not functioning well, the car as deliberating consultant and partner in the general negotiation of a lifestyle, something – or someone: at this point there is no longer any difference – with which you are connected. The fundamental issue becomes the communication with the car itself, a perpetual test of the subject's presence with his own objects, an uninterrupted interface.

J. BAUDRILLARD, THE ECSTASY OF COMMUNICATION

Not all work in this area closely mimics human and animal behaviour. SATORI TV **(FIG. 2.8)**, a small television that turns it head to face the viewer when touched, is one of the few objects designed at Cranbrook during the 1980s that goes beyond visual semiotics by using performance. This television suggests a life where our only company will be electronic

FIG. 2.7

In **THIRD HAND**, Stelarc wrote single words with a third artificial hand strapped to one of his own, activated by the EMG signals of the abdominal and leg muscles, while the real arm was remote controlled and jerked into action by two muscle stimulators.

domestic appliances, which must supply the missing banalities of everyday human contact. The artist Alan Rath goes one step further and literally gives technology a face, but not comfortingly. His faces are juxtaposed and recombined with other body and machine parts to create strange and sinister hybrids of people and machines. He uses videos of parts of the face, or whole faces held captive within cathode ray tubes: in c-clamp a face grimaces while its CRT container is held in a C-Clamp **(FIG. 2.8)**. Many of his pieces rely on puns, are comic and anthropomorphic, and remind us of our fear that machines might have lives of their own. But although such works remind us of a possible future where the human soul becomes literally trapped within the machine, their easy appeal means they are also easily forgotten; they are not disturbing enough to shock.

ALIENS

A range of possibility exists between ideas of the 'pet' and the 'alien'. While the pet offers familiarity, affection, submission and intimacy, the alien is the pet's opposite, misunderstood and ostracised. The artist, Martin Spanjaard explores this space, believing that:

In order to get used to talking to a machine, one should have one as a pet. A machine which has no particular function, and cannot actually be operated, but which responds to the events in its environment by producing spoken language. Like a cat, which rubs its head against you and meows when it wants to eat or go outside, or a dog which whines when you kick it.

D. VAN WEELDEN, **MACHINE VOICES**

Spanjaard's robot ADELBRECHT evolved over ten years, starting in 1982, from his desire to build a ball which would roll of its own accord and, when it collided with other objects and reverse, change direction or take other appropriate action. As technology developed so did ADELBRECHT; he can now sense whether he is being picked up or stroked, and whether and by how much light and sound are present, influencing his mood or 'lust' as it is termed by the artist. ADELBRECHT expresses the level of his 'lust' by rolling about and by a voice provided by the Institute for Research on Perception in Eindhoven. For example, if he has not been touched since becoming active, on becoming stuck he will call for help; but if he

FIG. 2.8
Peter Stathis' **SATORI TV** (1988),
which turns its head to face the viewer
when touched, suggests a life where
our only company will be the electronic
appliances of the home, which must
supply the missing banalities of
everyday human contact.

has been touched, he will call his owner. He says "Nice" on being stroked, and "Is it you?" on being picked up. The artist does not program ADELBRECHT to totally replicate human or animal psychology, which results in unexpected and quite poetic mumblings. ADELBRECHT is an example, as boundaries blur between ourselves and our digital environment, of where a new sense of 'alienation' or distance may be discovered. The electronic object does not have to fulfil our expectations, it can surprise and provoke. But, to fulfil this potential, designers need to leave behind a desire to model the new world of electronic products in their own, human, image.

(IN)HUMAN FACTORS

If user-friendliness characterises the relationship between the user and the optimal object, user-unfriendliness then, a form of gentle provocation, could characterise the post-optimal object. The emphasis shifts from optimising the fit between people and electronic objects through transparent communication, to providing aesthetic experiences through the electronic objects themselves.

But if aliens and user-unfriendliness are to be the alternatives to pets and user-friendliness, this user-unfriendliness does not have to mean user-hostility. Constructive user-unfriendliness already exists in poetry:

The poetic function of language has as its effect that when we read literature we become more aware of language than we are when we are confronted by language in its other functions. To introduce another term dear to the formalists, in literature language is 'foregrounded'. This, as Jakobson stresses, is the tendency of literature, much more fully recognised in poetry than it is in prose. In the everyday use of language it will seldom be practical and may even be found impolite to 'foreground' language. Everyday language is usually informative and instrumental; there is no call for either the speaker/writer or hearer/reader to dwell on the form of what is said/written since if a piece of information has been successfully passed or some action successfully instigated, the words by which this has been managed can count as 'transparent'. With the poetic function comes a certain opacity, for the writer is no longer passing information nor seeking to instigate action. There may also come an intentional ambiguity.

J. STURROCK, STRUCTURALISM

DEFAMILIARISATION.

The poetic can offer more than simply enriched involvement. It can provide a complex experience, critical and subversive. The Russian formalist poets of the 1920s based their ideology on estrangement. According to Viktor Shklovsky, the movement's best known exponent, the function of poetic art is to counteract the familiarisation encouraged by routine modes of

FIG. 2.9
Alan Rath's **C - CLAMP** (1992) literally gives technology a face, but not in a
comforting way. His faces are juxtaposed and recombined with other body and
machine parts to create strange and sinister hybrids of people and machines.

perception. We readily cease to 'see' the world we live in, and become anaesthetised to
its distinctive features.

Lebbeus Woods, an architect who has produced imaginary schemes (for instance, ORIGINS)
exploring this quality in building, refers to this strangeness as "objectivity", meaning not
an analytical state of mind but simply the appreciation of the objects as themselves,
independent of the operations of the mind upon them.

The effect of strangeness, infusing an encounter with the unfamiliar and the unknown, was used
by Bertolt Brecht to alienate the audience and make them aware that the institutions
and social formulae they inherit are not eternal and natural but historical, man-made,
and so capable of change through human action. He termed it the "A-Effect", developing
the conditions for informed appreciation rather than unthinking assimilation.
And Theodore Adorno wrote that authentic art could only function to resist totalisation if
it was strange and unfamiliar.

DESIGN AS TEXT

Despite an interest in linguistics and texts, the Cranbrook work stopped short of realising the full
potential of the model of meaning it pursued. Rather than radical provocations, it
produced beautiful, affirmative designs that were in literary terms structuralist rather
than post-structuralist.

Daniel Weil's work, on the other hand, shows what can be achieved if the notion of object as text
is taken to its (apparently illogical) logical conclusion, echoing the 'death of the author' in
literature. His designs challenge the observer to participate in constructing their meaning,
their questions, interpretations and criticisms becoming part of its meaning.

Weil's designs could be defined as a 'text' in Roland Barthes' definition: a 'space' of chains and
layers of meaning between the object and the viewer, continuously expanding with no
fixed origin or closure. When the boundaries of the work are demolished, the text opens
out onto other texts. Barthes redefined 'text', as a meta-linguistic mechanism that
reorganises the linguistic order, affecting the relationship between writing and reading.
Writing and reading, the pre- and post-textual, are of equal value and both writer and
reader are required to exert an equal effort of imagination. Similarly, in the case of a

FIG. 2.10
Daniel Weil's FOUR BOXES AND
ONE RADIO (1983) is a literal expression
of the fact that materials used in the design
of cases for radios have little intrinsic value,
but acquire value through the authorship
of the designer.

FIG. 2.11
Daniel Weil's SMALL DOOR (1986) is more obscure.
It challenges the viewer to participate in constructing
its meaning. The viewer's questions, interpretations
and criticisms are part of the object's meaning.

design object as text, designer and viewer play equal roles. This approach lends itself easily to electronic products, because their components can be freely arranged, unlike mechanical products where the arrangement of components is determined by technical constraints:

In Weil's view the object has a conceptual story which the person owning it has to complete... his approach is heavily influenced by Duchamp's conception of the 'unfinished picture'... for computer designers, as for Duchamp, the focus of their work now is the process of use of computer systems... security is not the objective. He offers a degree of understanding of technology, but control and domination over it are not assured.

J. THACKARA, THE CACOPHONOUS SOUNDS OF TECHNOLOGY

Weil's radios and clocks of the early 1980s are a good example of a research project exploring the aesthetic nature of electronic objects. Most products from this phase of his work seem transient and cheap. Thackara suggests this is an essential part of their nature, as their frailty reminds us of the delicate nature of our conceptual models too. They are objects about objects in the age of electronics, and express our changing relationship to objects brought about by electronic technologies. They sometimes do this clearly, as in FOUR BOXES AND ONE RADIO **(FIG. 2.10)**, a literal expression of the fact that all radios are packages in a box: the materials have little intrinsic value but acquire value through the authorship of the designer. At other times they do so more obscurely, as in SMALL DOOR, another design for a radio **(FIG. 2.11)**. Weil's designs are conceptual and open-ended, and challenge the user or viewer to engage with them. In literary terms they are post-structuralist.

Like most experimental designs for electronic objects during the 1980s, though, Weil's designs are reinterpretations of existing objects, primarily radios. Perhaps the radio is the electronic equivalent of the chair: a familiar and culturally rich object used by architects and designers as a vehicle to communicate new ideas. Although clocks and radios might seem trivial as technological objects, this is often the only level at which experimental electronic objects can be batch-produced without large investment. Ultimately, the radicalness of Weil's objects lies in their novel imagery and his open-ended approach to meaning. But they still package technology as a visual sign.

FIG. 2.12
Kei'ichi Irie's **LASCAUX CHAIR** (1988) began as a
design for a computer program. The structure of a practical
chair is a main routine; the program generates variants,
splitting legs in two, twisting and stretching elements.

BYPASSING THE SELF

Whereas the apparent strangeness of Weil's objects relates to linguistics and notions of the object as text, the architect Kei'ichi Irie and computer artist Masaki Fujihata use technology to give strangeness to non-technological objects. They explore ways of incorporating technology into processes that bypass our desire to model reality in our own image. The resulting artefacts are sophisticated and subtle fusions of what is and what might be. They map the interface between the social consciousness of the individual designer and the collective scientific consciousness, the dominant ideology embodied within the operating systems of the computer.

As a designer operating in a media-saturated cultural sphere, Irie utilises computer errors to escape making uncritical and unconscious use of existing cultural forms and conventions, and reproducing the ideology they encode. He considers designing to be autogenerative, made up of sub-routines. For Irie, when anything is possible, design is no longer about necessity but becomes a play between sub-routines, exploring what can be used rather than realising an optimum fit. A valid decision may be made on a whim for, as with Weil, the experience of the work is partly what the viewer brings to it:

Even in my own house at Sangubashi, the meaning came from the programming. Which is to say, the elements and methods I employed may have dictated a 70s Tokyo house, but that filter aside, you can see it was just a program. The final form did not have to come out like that at all. If I had applied another filter – who knows? – a tile roof might have resulted.

K. IRIE, COMPUTER CRASH BY DESIGN

His project for a chair **(FIG. 2.12)** experiments "with the interplay of noise and unadulterated parts". He first designed a computer program which generated different configurations for a chair with three legs and a seat. The structure of a practical chair is a main routine, but the program generates a host of variants, splitting legs into two, twisting and stretching elements. The designer simply edits, making selections and adjusting them to ensure they function as free-standing chairs. To Irie's delight, the addition of a number or two to the program can radically change the structure. He uses the computer as an extension of his consciousness: "My thought processes externalised in the form of a chair, which are in turn output as a terminal device 'chair'".

Irie applied this thinking to his work as an industrial designer with a large housing manufacturer. In his view each company has a "guiding will" programme or main routine. By understanding this programme it is possible to write 'bugs' into it, generating objects that are neither the familiar output of large corporations nor the singular expression of the designer as author, but a new, technologically-mediated collaboration between designer as virus and industry as program.

FIG. 2.13
In FORBIDDEN FRUITS (1991)
Masaki Fujihata regards these
computer graphic images as "virtual
fruit he is forbidden to hold".

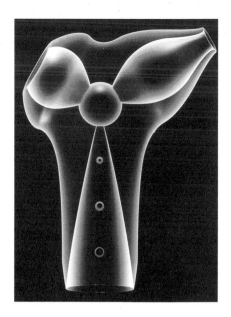

Fujihata responds to Tokyo's unique mixture of immaterial and material culture through an unconventional and conceptual form of industrial design. FORBIDDEN FRUITS realises computer visions **(FIG. 2.13)** using a CAD system designed for industrial designers and linked to a model-making system. An ultra-violet beam traces forms in a photosensitive resin which solidifies on contact with the light, creating translucent representations of computer data. His introduction claims that photography has generated a special "mental software" that is exploited by computer graphics. Interested in going beyond this to discover new potentials for computer graphics, Fujihata transports forms from the screen into the here-and-now, using a process very different from classical modes of making pictures and sculptures. He articulates data to edit form, using a tree structure to represent the process. On a whim, he returns to points, suddenly turning, constantly producing the tree map of his explorations from which grows "the virtual fruit he is forbidden to hold". (04)

FUNCTIONAL ESTRANGEMENT

The objects Irie and Fujihata produce focus attention on the design process. They do not challenge the way we experience reality. To provide conditions where users can be provoked to reflect on their everyday experience of electronic objects, it is necessary to go beyond forms of estrangement grounded in the visual and instead explore the 'aesthetics of use' grounded in functionality, turning to a form of strangeness that lends the object a purposefulness. This engages the viewer or user very differently than the relatively arbitrary results of Irie or Fujihata, the crude interpretations and explanations offered through the well-mannered and facile metaphors of mainstream design, or the soft cybernetics of the Human Factors community. This strangeness is found in the category of 'gadget' that includes antique scientific instruments and philosophical toys. Objects which self-consciously embody theories and ideas.

The fit between ideas and things, particularly where an abstract idea dominates practicality, allows design to be a form of discourse, resulting in poetic inventions that, by challenging laws (physical, social or political) rather than affirming them, take on a critical function. Such electronic objects would be conceptual tools operating through a language of

functionality which is entangled in a web of cultural and social systems that go beyond appearance.

Although transparency might improve efficiency and performance, it limits the potential richness of our engagement with the emerging electronic environment and encourages unthinking assimilation of the ideologies embedded in electronic objects. Instead, the distance between ourselves and the environment of electronic objects might be 'poeticised' to encourage sceptical sensitivity to the values and ideas this environment embodies. This could be done in a number of ways, of which the most promising is a form of functional estrangement: 'para-functionality'. This quality, common to certain types of gadget, is the subject of the next chapter which reviews projects and objects that work in this way, and explores how para-functionality could be applied to electronic objects.

N●TES

(01)
Ettore Sottsass quoted in P. Sparke, ETTORE SOTTSASS JNR, p. 63.

(02)
For an excellent critique of product semantics, see A. Richardson, "The Death of the Designer", DESIGN ISSUES (1993) 9 (2).

(03)
For a summary of John Dewey's views on aesthetic experience in terms of recognition and perception see page 59 below.

(04)
Brie and Fujihata's approaches superficially resemble that of John Frazer who has been involved in computer-generated form and structure since the early 1960s. Like many explorations of autogenerative models, especially in the field of artificial life, Frazer's inventions rarely move beyond the screen into physical space, although their formation often responds to data, such as environmental conditions, from sources outside the computer. See J. Frazer, THEMES VII: AN EVOLUTIONARY ARCHITECTURE.

This chapter reviews projects from art, architecture, and design which exemplify the functional estrangement I call 'para-functionality'. The term means here a form of design where function is used to encourage reflection on how electronic products condition our behaviour. The prefix 'para-' suggests that such design is within the realms of utility but attempts to go beyond conventional definitions of functionalism to include the poetic.

03
PARA-FUNCTIONALITY:
THE AESTHETICS OF USE

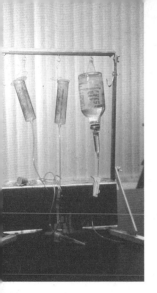

FIG. 3.1

Jack Kevorkian's **SUICIDE MACHINE** is a powerful piece of 'unofficial' design and shows how an industrial object can embody complex ideas through invention as a form of social criticism.

FIG. 3.2

This DRINKING CANE from the Saint-Etienne mail-order catalogue of 1910 operates in a context where etiquette assumes such importance that the object must be made to maintain it in a 'socially dangerous' situation.

ECCENTRIC OBJECTS: PARA-FUNCTIONALITY AND NON-DESIGN

Some naive, curious or eccentric objects, outside the world of conventional design, unintentionally embody provocative or poetic qualities that most product designs, even those intended to provoke, seldom achieve. Although industrial designers play a part in designing instruments of death (weapons) and pleasure (sex aids) these extreme areas of material culture rarely enter design discourse. Yet Jack Kevorkian's SUICIDE MACHINE, a powerful 'unofficial' design which materialises complex issues of law, ethics and self-determination, shows how an industrial invention can be a form of criticism **(FIG. 3.1)**. Critical of a legal system that outlaws euthanasia, Kevorkian has his machine to overcome this. Its ambiguous status between prototype and product makes it more disturbing than pure artworks by blurring boundaries between the everydayness of industrial production and the fictional world of ideas. It suggests a role for design objects as discourse where functionality can be used to criticise the limits which products impose on our actions.

At the other extreme is the world of antique walking sticks. A DRINKING CANE, designed for an alcohol merchant who must spend much of his time visiting the bars of his customers, discretely siphons off his drink while his host is not looking; a trigger later releases the drink into a gutter **(FIG. 3.2)**. It satisfies etiquette and exploits the walking stick's inherent potential for connection to other objects and contexts: hand, bar, glass, and gutter.

Walking sticks that become a card-table or seat **(FIGS. 3.3-4)** show how simple portable props can transform architectural spaces. They conceptually colonise the functional possibilities of pre-existing spaces. The user becomes a protagonist in a new narrative where a lobby or park becomes a casino. (01)

A third device, used by detectives in the 1940s for protecting fingerprints on a steering wheel, is beautifully absurd and surreal **(FIG. 3.5)**. Sigmund Freud cites G. Heymans' explanation that a joke works through bewilderment succeeded by illumination. (02) The word that is the vehicle of a joke often appears at first to be wrongly constructed, unintelligible, incomprehensible or puzzling. In this double steering wheel a similar unintelligibility is evident: its comic effect is produced by solving this bewilderment by understanding its function. This is also the case with 'Chindogu' **(FIG. 3.6)**. Their individual elements are recognisable, but the reason for combining them is at first bewildering.

FIGS. 3.3-4
The T A B L E C A N E, patented in England in 1891, and 'low seat cane' are examples of how simple portable props can transform an architectural space.

FIG. 3.5
A steering wheel, used by detectives during the 1940s to drive recovered vehicles back to the police station without smudging the thief's fingerprints.

The meaning behind the object is derived from 'sense-fiction': the objects make functional sense, but are still useless.(03)

FORBIDDEN EMOTIONS: PARA-FUNCTIONALITY AND DESIGN

In a review of an exhibition of work by Intermediate Unit 3, OBJECTS IN THE LANDSCAPE, at London's Architectural Association, Irie contrasts the "electronic devices essential to contemporary urban existence", the means whereby "information, entertainment and fantasy are promoted – and controlled", with the unit's "virus-like prototypes" **(FIG. 3.7)** that "invade and disrupt such networks, and propel minds and bodies into a hectically deregulated world of fragments – fragments of ideals, of illusions, of sensory impressions". The use of strange inventions by architects is not uncommon and, although they have lost much of their potency through over-use, their deployment in this instance as "bizarre monsters", designed to challenge the banal reality supported by consumer durables, emphasises the need to identify how electronic products can offer alternative expressions of their own functional logic. In a field where "product design is thoroughly integrated in capitalist production, [and] bereft of an independent critical tradition on which to base an alternative",(04) only a few designers use the function of products as criticism.

For example, Penny Sparke cites Gaetano Pesce: **(FIG. 3.8)** his "use of distortion and exaggeration [are] 'absurd' devices for commenting upon his observations. Rather than turning to alternative media, Pesce uses the language of design to make its own self-commentary",(05) but his objects do not incorporate functionality as a primary component. When functionality does enter, it is often jokey and closer to the playful one-off multiples created by Fluxus. During the 1980s Denis Santachiara and Philip Garner developed approaches that merit a closer look. Santachiara, who developed a distinctive approach over many years, aims to raise the aesthetic quality of mass-produced everyday objects such as domestic appliances by developing their possibilities of animation. This could be seen as little more than a desire to use technology to give objects a personality by making them more expressive and quirky **(FIG. 3.9)**. But his concern is with an aesthetics of use which give objects a distinctive identity from the linguistics of construction and manufacture. Santachiara subverts technical knowledge, redirects it towards provocative

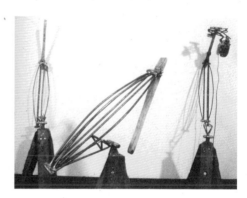

ends, provides more than enriched interactivity, and raises the complex issues of what Baudrillard has called the "crisis of functionalism".

Baudrillard argues that the acceptance of functionalism as an arbitrary but dominant rationality gave rise to an irrational counter-discourse which moves between the two poles of kitsch and surrealism:

> The surrealist object emerges at the same epoch as the functional object, as its derision and transgression. Although they are overtly dys- or para-functional, these phantasmic objects nevertheless presuppose – albeit in a contradictory sense – the advent of functionality as the universal moral law of the object, and the advent of this object itself, separated, autonomous and dedicated to the transparency of its function. When one ponders it, there is something unreal and almost surreal in the fact of reducing an object to its function: and it suffices to push this principle of functionality to the limit to make its absurdity emerge. This is evident in the case of the toaster, iron or 'undiscoverable objects' of Carelman.

> J. BAUDRILLARD, FOR A CRITIQUE OF THE POLITICAL ECONOMY OF THE SIGN

Santachiara's work is often closer to kitsch than Garner, who is closer to surrealism and the absurd. Garner's proposals for products are a form of industrial design that taps into the strange psychological and social narratives arising from the objects themselves and interaction with and through them in a consumer-oriented society. Although their overtly satirical and whimsical character, often simply visual puns or jokes, undermines the viewer's suspension of disbelief **(FIG. 3.10)**, they demonstrate the power of mock-ups, scenarios and fictitious narrative over working prototypes as a way of presenting this kind of fiction. The success of both his books confirms that people understand and relate to the narrative behind the work without having to use the objects.

Santachiara and Garner operate within the realm of the gadget, the opposite of the well-designed object. The term 'gadget' here denotes a curious, original and witty accessory of no real use, as opposed to the 'gimmick' which is too transparent in its effort to impress and attract attention. Giulio Ceppi remarks that "probably the gadget has never been considered, by official design culture, as the result of a design effort, an industrial product capable of revealing

FIG. 3.8
Gaetano Pesce's furniture for Cassina during the early 1960s uses
the language of design to communicate his observation that people
will always be alienated from objects as long as consumption is
the primary reason for an object's existence.

interesting technical features or of influencing peoples behaviour" and that:

The most important phenomenon caused by the gadget is, however, a psycho-behavioural
factor: wonder... The fact that wonder and surprise are two variables that rarely enter into the
design of industrial objects has induced the development of a clandestine niche in which such
forbidden emotions can be found.

G. CEPPI, PLAYING WITH TECHNOLOGY

HETEROTOPIAN GADGETS: PARA-FUNCTIONALITY AND ART OBJECTS

For examples that explore the aesthetics of this "clandestine niche" of forbidden emotions it is
necessary again to move away from industrial design, and begin with literature: not the
gadget-ridden world of science-fiction but a world where writing itself is a gadget in that
it celebrates the workings of language. The heterotopia described by Michel Foucault
illustrates what a literary gadget might be like:

Utopias afford consolation: although they have no real locality there is nevertheless a
fantastic, untroubled region in which they are able to unfold; they open up cities with vast
avenues, superbly planted gardens, countries where life is easy, even though the road to them
is chimerical. Heterotopias are disturbing, probably because they destroy 'syntax' in advance,
and not only the syntax which causes words and things (next to and also one another) to 'hold
together'. This is why utopias permit fables and discourse: they run with the very grain of
language and are part of the fundamental dimension of the fabulous; heterotopias
(such as those found so often in Borges) desiccate speech, stop words in their tracks,
contest the very possibility of grammar at its source; they dissolve our myths and sterilise the
lyricism of our sentences.

M. FOUCAULT, THE ORDER OF THINGS

David Porush, uses terminology that invites comparison between the poetics of real machines and
strange inventions, and literary gadgets:

FIG. 3.9
Denis Santachiara's P O R T A L E (1989)
which sparks when it is passed
through, is an example of his concern
with an aesthetics of use where
invention is used to give objects
a distinctive identity that moves away
from the linguistics of construction
and manufacture.

FIG. 3.10
Philip Garner's A L I E N A T U R E (1985) demonstrates the power of
mock-ups, scenarios and fictitious narrative over working prototypes
as a way of presenting this kind of fiction.

[Samuel Beckett's] Lost Ones is a palpable fiction which, even as its inventor attempts to complete the blueprint, collapses into impossible meaninglessness, self-contradiction, and absurdity. The fallibility of the cylinder machine lies in the fact it is constructed in words; the author's attempt to describe it precisely becomes an exercise in the futility of trying to describe anything using language.

D. PORUSH, THE SOFT MACHINE

Beckett uses two kinds of language, a precise technical/mathematical one, and a language of "failure, probability and doubt". These two rhetorics are at odds with each other and their weaving together provides the qualities of this text, "an allegorical world of pure fiction" about the "perception of the mute resistance of worldly objects to our vain and inappropriate attempts to attach names to them". Paul Klee seems to have incorporated this sensibility into his drawings: for example THE TWITTERING MACHINE **(FIG. 3.11)**, where a strange device hovers in the imaginary space of the drawing, suggesting a realm where machines do not simply mirror rationality through nonsensical functions but embody alternative physical laws to ours, like Marcel Duchamp's 'Large Glass' and the 'Pataphysics' of Alfred Jarry.

What happens when this sensibility moves from the page and canvas to become part of everyday space? The sculptor Panamarenko is interesting in this respect as his machines embody the same ambiguity as the literary and painterly gadgets of Beckett and Klee. Whereas artists like Jean Tinguely have constructed useless machines that comically mirror rationality, Panamarenko's objects rarely work **(FIG. 3.12)**, provoking the viewer to think about the nature of invention and the desires that motivate it. They are about flight, desire, the limits of knowledge and the transition from wondering and dreaming to the dull reality of realisation. By denying that last step and conventional practice they hover successfully between the imaginary and the real. His scientific theories on flight also highlight the fictional nature of scientific knowledge and blur the boundaries between words and things.

FIG. 3.11
Paul Klee's **THE TWITTERING MACHINE**
(1922) shows a strange device hovering in the imaginary space of the drawing, suggesting a realm where machines do not simply mirror rationality through nonsensical functions.
PAUL KLEE, THE TWITTERING MACHINE (1922). COPYRIGHT DACS 1999.

The inventor-artist Steven Pippin meditates on photography. He coats with photographic chemicals the interior surfaces of everyday objects like washing machines, toilets and bath tubs, turning them into cameras. His ingenious experiments interweave the host object's original functionality with that of a camera, resulting in objects that occupy a difficult conceptual space outside the usual polarisation of functionalism and surrealism. They do produce sense, and we understand them, but it is hard to say what exactly we understand about them. They differ from the symbolic machines and devices of Rebecca Horn **(FIG. 3.13)**, where things do what we expect but the company they keep surprises. Pippin creates conceptual gadgets that render useless our expectation of what things ought to do; they turn knowledge itself into a gadget and allow us to catch glimpses of how knowledge works and wonder at its beautiful but useless mechanisms.

The objects produced by inventor-artist Philippe Ramette occupy a different part of the space between ideas and things. They resemble in atmosphere the design proposals of Philip Garner but are less ironical in their straightforward presentation of function through the nostalgic language of antique scientific instruments. Meyer Rubinstein[06] describes them as "prostheses of the spirit", aids to thought and contemplation. As with many of the objects described in this chapter, the emphasis on functionality focuses the viewer's attention on the space between the experience of looking at the work and the prospect of using them. Here the emphasis is on the body and its relationship through the senses to the space that contains it. Although fully working, many of Ramette's objects cannot be used because they can hurt or worse: for example OBJECT TO MAKE YOURSELF BE STRUCK BY LIGHTNING, or INTOLERABLE OBJECT whose lens focuses sunlight onto the top of the head. But not all his objects are threatening. In a world of artificial objects shaped almost entirely by functionalism, devices like an OBJECT WITH WHICH TO SEE THE WORLD IN DETAIL do not attempt to escape the dictates of functionalism but instead work from within, extending it to include the poetic and playfully subversive **(FIG. 3.14)**.

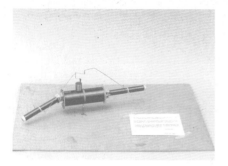

FIG. 3.12
Panamarenko's VOYAGE TO THE STARS (1979) like many of his
other pieces, does not actually work. This provokes the viewer to think
about the nature of invention and the desires that motivate it.

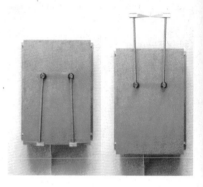

FIG. 3.13
The UNCONSCIOUSNESS OF FEELINGS (1983)
by Rebecca Horn is a symbolic machine where things do
what we expect, but the company they keep surprises.
REBECCA HORN, THE UNCONSCIOUSNESS OF FEELINGS (1983),
COPYRIGHT DACS 1999.

SOCIAL-FICTIONS: PARA-FUNCTIONALITY AND CRITICISM

Although often threatening, Ramette's objects do not shock. Their critical content is hidden
beneath the poetry of construction and the humorous appreciation of their function. But
the work of the artist Andrea Zittel shocks by using the familiar contexts of the home, and
of the system of production and consumption, to concretise alternative values that are
outside notions of the future or past but sit uncomfortably alongside 'now'. They suggest
that the way things are may not be the only possibility. They initiate a questioning and
awareness that helps unravel the 'one-dimensionality' that characterises present times
and maintains 'the impossibility of the possible'.(07) Zittel's COMFORT UNITS suggest an
unusual way of thinking about the role of furniture **(FIG. 3.15)**. Her emphasis shifts from
style and image to their psychological use as tools for inhabitation. By clearly favouring
the manifestation and fusion of particular functional possibilities over others they remind
us, through an extreme but credible form of functional reductionism, of our dependence
on objects for developing new behaviours. In her work it is never quite clear whether her
positive-reinforcement prototypes reflect a genuine belief that this is what we need, or
are an ironical play on modernism.

The architects Kenneth Kaplan and Ted Krueger (K/K Research and Development) leave no doubt
about the status of their assemblages of found machine parts **(FIG. 3.16)** as ironic
'analogues' for architectural ideas. Although their writing is polemical their use of objects
to attract the attention of the audience, before it is seduced by their usually written
political narrative, reduces the objects to dumb props. Their CRIB-BATIC project (with
Christopher Scholz), however, is an exception **(FIG. 3.17)**. A prototype for a child's push-
chair made from steel (they felt children needed to be exposed to hard materials from an
early age), it was equipped with measuring equipment so that the child might interact with
the environment on the go. This piece is more powerful than their more obtuse
architectural analogues, because it is possible to imagine what it would mean for such
thinking to enter everyday life through similar objects. It moves beyond implied
functionality and appearances to use function to draw attention to the role objects play
in conditioning our responses to the environment.

FIG. 3.15
Andrea Zittel's C O M F O R T U N I T S (1994) suggest an
unusual way of thinking about the role of furniture.
Her emphasis shifts from issues of style and image to
their psychological use as tools for inhabitation.

FIG. 3.14
The emphasis placed on functionality in Philippe Ramette's
O B J E C T W I T H W H I C H T O S E E T H E W O R L D
I N D E T A I L (1990) focuses the viewer's attention on the
space between the experience of looking at the work and
the prospect of using it.

Another architectural practice, Diller + Scofidio, designs and builds architectural gadgets that work on a critical level. PARA-SITE, an architectural exploration of the impact of electronic media on architectural space, is relevant here because of the equal importance it gives to electronic and conventional media. Electronic objects such as televisions and video cameras are not repackaged or redesigned but integrated into new hybrid objects **(FIG. 3.18)**, transforming these boring and familiar devices into an architectural intervention. Diller + Scofidio deploy technology intelligently, using it to reveal, enable and criticise, intervening in not only the abstract space of the building but also its social and practical use.

PARA-SITE is one of many critical interventions in public spaces by architects and artists. One of the best known is Krzysztof Wodiczko's large scale projections onto public buildings. He has written: "My socio-aesthetic research and experiences deal with 'strategies' for making public art critical, non-official art". He studied on the graduate programme of industrial design at the Akademie Sztuk Pieknych in Warsaw under a former collaborator of Le Corbusier, Jerzy Soltan, who advocated a '(post)-avant-garde' strategy of critical engagement with and infiltration of, the institutional structures of industry and culture. On graduation he worked in Warsaw as an industrial designer for UNITRA, a manufacturer of electronic products. One of his first pieces of art was done in 1969 while still an industrial designer there: PERSONAL INSTRUMENT(08) **(FIG. 3.19)**. He was assisted in this by technicians from the Experimental Music Studio in Warsaw:

The instrument transforms the sounds of the environment.

The instrument functions in response to hand movements.

The instrument reacts to sunlight.

The instrument is portable.

The instrument can be used any place and any time.

The instrument is for the exclusive use of the artist who created it.

The instrument permits him to attain virtuosity.

K. WODICZKO, INSTRUMENTS, PROJECTIONS, VEHICLES

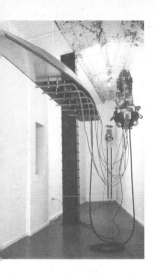

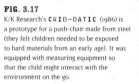

FIG. 3.17
K/K Research's **C R I B – B A T I C** (1986) is
a prototype for a push-chair made from steel
(they felt children needed to be exposed
to hard materials from an early age). It was
equipped with measuring equipment so
that the child might interact with the
environment on the go.

FIG. 3.16
K/K Research's BUREAU–DICTO
(1989) is an ironic 'analogue' for
architectural ideas consisting of an
assemblage of found machine parts.

Wodiczko has said that "the instrument's magic silence is its socio-political message". Although private, it depends on a public space as a source of sound, and so that others can gaze at it and imagine how it works. "It was a way to shape a metaphor for the limits to the freedom of the individual Pole, to the ways he could exercise this freedom, and to his power in relation to public spaces". It was not designed for mass production nor even for a limited edition "and yet it was intended for the whole world as a metaphor for community life and the nature of public spaces in Poland":

My personal instrument proved to be the point of departure for all my public works. it was my first attempt to provide a metaphorical definition of man's position as a 'citizen' of a dominated public space. It was also the first time I attempted to hint at the 'strategy' of taking words and using space as medium in which to speak them, even though the right to use a private voice in space that was totally 'socialised' (politicised) by the government was utterly non-existent. I proposed the technique of speaking silently, reticently or by grotesquely exhalting silence.

K. WODICZKO, INSTRUMENTS, PROJECTIONS, VEHICLES

Wodiczko's public projections and homeless vehicles continue this research **(FIG. 3.20)**. A less known object, the **ALIEN STAFF**, shows how industrial design, through conceiving new functions and their configuration as 'accessible' products, can function critically. The staff houses a small LCD television, while a small video player, a CB radio or walkie-talkie, and batteries are in a shoulder bag. The small size of the display, its position at eye level and its proximity to the alien's face are all important. Once somebody has been attracted, a relationship is perceived between the face within the screen and the actual face of the alien, conceptual barriers are destabilised, and real communication may begin:

It is an instrument that gives the individual immigrant a chance to 'address' directly anyone in the city who may be attracted by the symbolic form of the equipment and the character of the 'broadcast' program.

K. WODICZKO, INSTRUMENTS, PROJECTIONS, VEHICLES

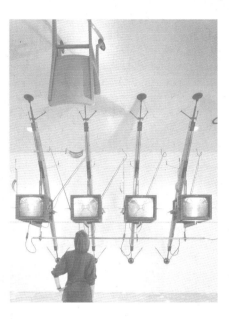

FIG. 3.18
Diller + Scofidio's P A R A – S I T E (1989), an architectural exploration of
the impact of electronic media on architectural space, gives equal
importance to electronic and conventional media. Electronic objects such
as TVs and video cameras are not repackaged or redesigned but are integrated
into new hybrid objects.

Wodiczko's designs show how simple electronic technologies can challenge preconceptions, but are at the margins of design. Although I see them as design proposals not artworks it seems that, to hold a design view where electronic objects function as criticism, one must move closer to the world of fine art because the design profession finds it difficult to accommodate such research. Objects such as PERSONAL INSTRUMENT and ALIEN STAFF, with their use of simple electronics and their emphasis on invention and social and cultural content, are rare examples of how product design and the electronic object can fuse into critical design.

HERTZIAN PATHOLOGIES:
PARA-FUNCTIONALITY AND ELECTRONIC OBJECTS

People like to play lotto and people like to use the ATM. Why don't you make it an option in the ATM to say put your money in and say, I'll bet a little bit and see if I can get a little more out, so you ask for twenty dollars, and you push the button, and you could get twenty-five or you could get fifteen.

JEFF KIPNIS, IN OFF RAMP

Another zone of activity outside that of even the exiled designer is 'anonymous design', where alternative conceptual models already find expression through electronic artefacts. 'Pathological' gadgets are examples of life outside the normal conception of reality; they are design fictions, deviations and failures and help to maintain the 'impossibility of the possible'.

Many of these devices concern communication. Most communication technology is oriented towards the individual; it cannot yet support or even encourage more complex social situations. It is point-to-point, one-to-one, not place-to-place. Yet most of this narrow form of communication takes place within that vast field of telematic possibility, the electromagnetic spectrum. The tools and devices limit the possibilities, not the medium. Ironically, many of the more interesting possibilities can be found in 'pathological' products based on paranoia and suspicion. Many are designed to 'open up' one-to-one channels, transforming private situations into public ones. Scanners, bugs, and detectors

FIG. 3.20
Krzysztof Wodiczko's **HOMELESS VEHICLE** (1988-89).

FIG. 3.19
Krzysztof Wodiczko's **PERSONAL
INSTRUMENT** (1969), although
private, depends on a public space
as a source of sound, and so that
others can gaze at it and imagine
how it works.

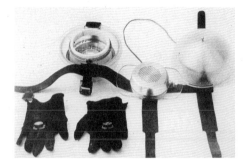

illegally 'socialise' the world of private telematics. For example, scanners have tuned into wireless baby intercoms enabling 'recreational voyeurs' to listen into intimate bedroom conversation.

The radio scanner[09] hovers at the limits of legality **(FIG. 3.21)**. In the UK it is legal to make and sell it but, like many pieces of surveillance equipment, not to use it for eavesdropping. It draws attention to what DeLanda has termed the "policing of the spectrum", not a public space but a highly policed and militarised state space. It is one thing to be prosecuted for eavesdropping but, if the information is passed on to a third party and worse, sold, it becomes a serious offence. If sensitive frequencies are found stored in the memory, the owner is likely to be prosecuted. That the radio scanner is a powerful object, entangled with the social and legal systems of society, has been recognised by the artist and musician Robin Rimbaud alias Scanner:

To Scanner, the world of the personal phone call – an easily tapped medium, especially if you've been building your own radio sets since your teens – represents a far more honest depiction of the world than the outpourings of televisual reality. And Scanner's records, packed with a huge collection of telephone 'normality', are, in turn, far more real and disturbing than any arty fabrication of reality.

S. KING, IN TRANCE EUROPE EXPRESS 2

The radio scanner enables new urban maps to be made, revealing normally hidden structures of the visible and conventional. The scanner is a meta-radio: it transcends the many categories of radio incorporated into commodities, highlighting their commonality as parts of an electromagnetic spectrum.

The DR. GAUSS EMF detector is one step further up the evolutionary ladder of gadgets **(FIG. 3.22)**, a low-cost version of a usually expensive piece of equipment, used to measure the magnetic component of possibly harmful electromagnetic fields in the home. The device is simply a black box, but the act of using it reveals its conceptual power: when it picks up a field it screams, rising in pitch with the strength of the field.

Objects like this allow us to develop new conceptual models about our environment, helping us to see invisible structures and patterns. They often occupy the cultural wasteland of in-flight

FIG. 3.21
The scanner is an example
of a 'pathological product' based
on suspicion and paranoia designed
to open up one-to-one channels,
transforming private situations
into public ones.

FIG. 3.22
The DR.GAUSS EMF
detector allows the owner
to gather information about
the presence of harmful
electromagnetic fields so that
a complaint can be made.

magazines, Sunday supplements and specialist shops, where alternative world views embodied as material reality exist as a non-serious and marginal phenomena. But in showrooms they become vital alternatives to art works and galleries. Whereas people step out of ordinary life into an art gallery, the contents of showrooms relate directly to everyday life in the mind of the window-shopper.

BETWEEN RATIONALITY AND REALITY

The most effective examples in this chapter function as test-pieces that, through their marginalisation, make visible the barriers limiting poetic experience in everyday life. The apparent unusability of many of these objects creates a heightened sense of 'distance'. This can be because the objects do not work technically or, because they are conceptually difficult to assimilate. To see that they are usable is to acknowledge that existing notions of functionality have been extended, a result of imagining uses for these objects. They challenge the impossibility of the possible. It is not enough to look and decode their visual iconography: they must be used. Through use, or at least by modelling a scenario of use in the mind, the observer discovers new ways of conceptualising reality. They dismantle conceptual models which limit the way we use artefactual reality to extend our scope for action. They challenge how we think about extensions to our 'selves' in ways that do not simply magnify but, rather, transform our perception and consciousness of our relation to our environment.

They share no coherent theory. They are simply stories, but stories that allow complex interactions between reality and imagination. Driven by poetry, imagination and intuition rather than reason and logic, they have their own rationality, an alternative to our everyday scientific-industrial rationality. These are stories about the space between rationality and reality, which in an industrial society have come to be synonymous. When these props are introduced into everyday life as a 'virus', subverting it, people can participate in the story, exploring the boundaries between what is and what might be. This is the role of the para-functional as criticism.

By imagining the object in use, we become lost in a space between desire and determinism. Within this space lies the bizarre world of the 'infra-ordinary', the subject of the next chapter, which reviews a number of projects in relation to behaviour and narrative.

N●TES

(0 1)
This also suggests a way of establishing an architectural role for the object in the sense of Bernard Tschumi's "there is no space without event, no architecture without programme; the meaning of architecture, its social relevance and its formal invention, cannot be dissociated from the events that 'happen' in it". B. Tschumi, "The Discourse of Events" in THEMES 3, p. 17.

(0 2)
S. Freud, JOKES AND THEIR RELATION TO THE UNCONSCIOUS, p. 13.

(0 3)
"The Japanese word 'Chindogu' literally means an odd or distorted tool — a faithful representation of a plan that doesn't quite cut the mustard... they are products that we believe we want — if not need — the minute we see them. They are gadgets that promise to give us something, and it is only at second or third glance that we realise that their gift is undone by that which they take away." K. Kawakami, 101 UNUSELESS JAPANESE INVENTIONS, pp. 6-7.

(0 4)
Thackara, DESIGN AFTER MODERNISM, p. 22.

(0 5)
Sparke, ETTORE SOTTSASS JNR, p. 52.

(0 6)
M. R. Rubenstein, "Phillippe Ramette" in ART FORUM May 1993, p.100.

(0 7)
I refer to those cultural mechanisms that marginalise alternatives to the present, even when economically and technically feasible, as utopian and 'unrealistic'.

(0 8)
This project is of personal interest to me because a similar project, the NOISEMAN (1989) marked my first experience of designing in a critical way while working for a consumer electronic company. My NOISEMAN, less self-consciously political than Wodiczko's work, focused on the role of mass-production and on creating new aesthetic experiences of urban spaces.

(0 9)
A wideband radio of the kind which, tuned into a mobile phone conversation between Princess Diana and a lover, recorded the 'Squidgy tapes', extracts of which were published in the tabloids in early 1990s.

'Psychosocial narratives' refers to the unique narrative potential of electronic products, the world of desire and fiction that embraces consumer goods, the socialisation which the use of electronic products encourage, and the idea that behaviour is a narrative experience arising from the interaction between our desire to act through products and the social and behavioural limitations imposed on us through the conceptual models they impose. For instance, although an essential part of everyday life, the telephone embodies crude concepts of social etiquette compared to furniture and architectural space.

This chapter looks at the following ideas: the user as a protagonist and co-producer of narrative experience rather than a passive consumer of a product's meaning; how the psychological dimensions of experiences offered through electronic products can be expanded to include darker conceptual models of need – usually limited to cinema and literature – by referring to the world of product misuse and abuse; the lack of work by authors and film-makers exploring this area, despite its prevalence in everyday life; the idea that the designer, in their role as a provider of new behavioural opportunities, becomes an 'author' working in a medium that can present experiences rather than represent them; and how the electronic product becomes a 'role model' bringing about transformations of perception (and conception) in the user as a protagonist by embodying unusual psychological needs and desires in 'pathological' electronic objects.

04
PSYCHOSOCIAL
NARRATIVES

USER AS PROTAGONIST

The phone and the film projector surely need us to bring them to life, to dial the number or to flick the switch, yet when these machines take us to places, people, and ideas otherwise forbidden, so they flick the switch on us. The beauty of new technology is that by engaging our imaginative co-operation it moves a flat two-dimensional relationship of subject/object, man/machine through a magical door into new mental territory.

C. BEEVOR, IN NATO

The narrative possibilities suggested in this quotation differ from the conventional narrative dimensions created for the electronic product by applying semiotics. Examples of the latter are the SPORTS WALKMAN produced by Sony in the mid-1980s which referred to imagined contexts of use, and the O-PRODUCT camera designed by Water Studio for Olympus, in both of which form and texture are manipulated to evoke a world of fantasy and fiction, blurring distinctions between everyday life and the hyper-reality of advertising and soap opera. ·

One way of viewing this difference can be found in THE MEANING OF THINGS which outlines an approach to aesthetic experience developed by Rochberg-Halton and based on Dewey's 1934 distinction between perception and recognition:

For Dewey, recognition describes a falling back on some previously formed interpretative schema or stereotype when confronted with an object, whereas perception involves an active receptivity to the object so that its qualities may modify previously formed habits or schemes. Although the explicit purpose of art is to evoke aesthetic experience, Dewey does not limit aesthetic experience to art alone but considers it a potential element of all experience. Perception is essential to aesthetic experience and leads to psychological growth and learning. Recognition, or the interpretation of an object or experience solely on the basis of already existing habits, only serves to condition a person further to a life of convention. If culture were simply a symbol system of convention, as some cognitive anthropologists argue, then aesthetic experience would only consist of recognition in Dewey's sense, because the object of that experience 'contains' meaning only as an arbitrary sign endowed with meaning by cultural convention and not because of unique qualities of its own.

CSIKSZENTMIHALYI & ROCHBERG-HALTON, THE MEANING OF THINGS

FIG. 4.1
This hand-held scanner/fax is an example of
an (ab)user-friendly product used to exploit
the potentially subversive possibilities of the
parallel world of illicit pleasures stolen from
commodified experience.

The narrative possibilities offered by the conventional semiotic-based approach depend on 'recognition', whereas the more dynamic form of narrative suggested by Beevor could open the way for the active critical receptivity of an experience that 'perception' involves.

In the case of electronic products the "unique qualities" of the object of interaction are their potential as an electronic product to persuade the users as protagonists through the user's use of the object, to generate a narrative space where the understanding of the experience is changed or enlarged. By using the object the protagonist enters a space between desire and determinism, a bizarre world of the 'infra-ordinary', where strange stories show that truth is indeed stranger than fiction, and that our conventional experience of everyday day life through electronic products is aesthetically impoverished.

THE INFRA-ORDINARY

The machine does what the human wants it to do, but by the same token the human puts into execution only what the machine has been programmed to do. The operator is working with virtuality: only apparently is the aim to obtain information or to communicate; the real purpose is to explore all the possibilities of a program, rather as a gambler seeks to exhaust the permutations in a game of chance. Consider the way the camera is used now. Its possibilities are no longer those of a subject who 'reflects' the world according to his personal vision; rather, they are the possibilities of the lens, as exploited by the object.

J. BAUDRILLARD, XEROX AND INFINITY

In 1994 the British mobile phone company Cellnet produced a booklet, MOBILE MOMENTS: A COLLECTION OF TALES FOR THE '90S, a chronicle of events which, it felt, demonstrated the crucial part the mobile phone has come to play in our lives. The tales are arranged under headings such as "Mating by Mobile", "Mobile Heroes", "Mobile Marvels" and, most interestingly for this chapter, "Mobile Mishaps". Each story is an example of the narrative space entered by using and misusing a simple electronic product, of how interaction with everyday electronic technologies can generate rich narratives that challenge the conformity of everyday life by short-circuiting our emotions and states of mind. I am recommending, not that designers try to predict misuses of products, but rather that they refer as a context of use to this rich narrative space instead of the models of normality usually referred to when new functional possibilities are being developed.

In my opinion, the really interesting relations between people don't occur in the form of communication. Something else happens: a form of challenge, seduction, or play which brings more intense things into being. By definition, communication simply brings about a relationship between things already in existence. It doesn't make things appear. And what is

FIG. 4.2
Nam June Paik's RANDOM ACCESS (1963) is an example
of a device where the artist has invented new interactive possibilities
for existing products that draw attention to the limitations imposed
by manufacturers through unimaginative design on our experience of
everyday electronic products.

more, it tries to establish an equilibrium – the message and all that. Yet it seems to me that there is a more exciting way of making things appear: not exactly communication, but something more of the order of challenge. I'm not sure that this would invite an aesthetic of communication strictly speaking.

J. BAUDRILLARD, THE REVENGE OF THE CRYSTAL

Some people already exploit the potentially subversive possibilities of this parallel world of illicit pleasures stolen from commodified experience. They seek out (ab)user-friendly products that lend themselves to imaginative possibilities for short-circuiting the combinatorial limits suggested by electronic products. In FEXY FACTS, Alfred Birnbaum writes about the abuse of hand-held scanners/fax machines **(FIG. 4.1)** by perverts to scan parts of their bodies through sheets of clear plastic and fax the resulting distorted images to lone women. Another example is Douglas Gordon, who appeared on television and used two telephones to call two galleries, fixing the phones together and recording the resulting surreal conversation.

These stories form part of a pathology of material culture that includes aberrations, transgressions and obsessions, the consequences of and motivations for the misuse of objects, and object malfunctions. This is related to the conceptual strategy explored by John Cage in RADIO MUSIC (1974) and his many pieces for prepared pianos, and by Nam June Paik in his MAGNETIC TV and RANDOM ACCESS **(FIG. 4.2)** audiotape wall. Both artists show behaviours towards technology that invite others to follow. Concerned with software not hardware, they invent new uses for existing technologies and promote interaction with 'designed' objects that subvert their anticipated uses. In doing so, they challenge the mechanisms that legitimise the conceptual models embodied in the design of the product or system (piano, television or tape machine). In his video MAKING DO AND GETTING BY (1995) the sculptor Richard Wentworth documents our natural ability to subvert object types and act in new ways on our environment. Often, as a by-product of trying to solve a practical problem, a poetic result is achieved, as different ideas, embodied in objects but usually kept apart, come together to reveal hidden similarities. Cartoons and comedies also present a world where the conventional use of

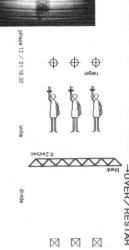

everyday objects is turned on its head, leading to surreal and of course humorous situations.

One of the Frankfurt School's arguments was that pleasure has been de-sublimated and is only available through buying consumer goods. When an object's use is subverted, it is as though the protagonist is cheating the system and deriving more pleasure than is his or her due. The subversion of function relates to a breakdown of order; something else becomes visible, unnameable, unable to find a correspondence in the material world. This subversion of function is related to not being able to find the right word, creating neologisms that bend language to accommodate something new. Desire leads to a subversion of the environment creating an opportunity to reconfigure it to suit our 'illegitimate' needs, establishing new and unofficial narratives.

Although the misuse and abuse of everyday objects is related to the anthropological study of material culture there is little literature on its surreal aspects. Occasional overlaps with urban legends (01) establish connections with anthropology proper, and offer at least some analysis of this subject, although emphasis is on collating, and discussing the truthfulness of, the legends themselves. In MYTHOLOGIES, Barthes famously introduced the role played by commodities in the formation and consumption of popular culture, but pre-dates the explosion of electronic products which shape nearly every aspect of modern life.

The almost unbelievable stories reported in tabloid newspapers, testify to the unpredictable potential of humans to establish new situations despite the constraints on everyday life imposed through electronic objects. A mother shoots her son after an argument over which television channel to watch; the police set a trap for scanner snoopers by broadcasting a message that an UFO has landed in a local forest (within minutes several cars arrive and their scanners are confiscated); a parent is outraged by a speaking doll, made in China, that appears to swear.

A PATHOLOGY OF MATERIAL CULTURE

The aesthetic potential of the narrative space centred on the consumer product has received surprisingly little attention from artists and writers and even less from designers. Few films or stories acknowledge how our lives and identities are intertwined with machines and artefacts, particularly everyday electronic products. Though we inhabit an environment of electronic gadgets and gizmos, little effort is turned towards exploring what this means. The film FAMILY VIEWING (1987) by the director Atom Egoyan studies the relationships between members of a family mediated through everyday technologies such as the telephone and the video recorder. Their sometimes unconventional use of these banal technologies is seamlessly integrated into their lives. This encouraging vision of technology, where new media allow additional forms of expression for everyday desires, offers an alternative to Hollywood's sci-fi and shock/horror visions of technology.

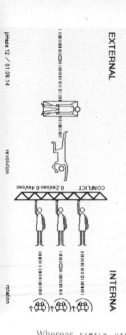

EXTERNAL

CONFLICT 0.2m/sec-0.4m/sec

INTERNAL

FIG. 4.3
Dumb Type's P H (1990) is one
of several performances they have
created in response to the impact
of consumer technology, through
commodification, on our lives.

Whereas FAMILY VIEWING focuses on technologically mediated relationships between people, Nicholson Baker's novels THE MEZZANINE, VOX, and ROOM TEMPERATURE richly exemplify how product-centred narratives can evolve from interactions with everyday objects. Deploying a refined appreciation of mass-produced material culture, he weaves playful narratives around psychological and physical interactions with and through the most banal artefacts, laying bare the usually hidden mechanisms of everyday material pleasures. In ROOM TEMPERATURE, for example, he writes about how as a child, by repeatedly cycling over an electronic-traffic counting cable, he might have influenced the future traffic system of his city. At the same time he exposes the failings of an over-scientific traffic-planning method based on mathematical models rather than "non-mathematical acts of judgement based simply on years of driving indignantly around".

Dumb Type, a group of artists, writers and architects based in Kyoto, have developed performances about the cultural and behavioural aspects of consumer technology, and commodification's impact on our lives. The performance PH **(FIG. 4.3)**, staged in a pit below the audience, involves visual projections, sound, and dance. A special piece of stage machinery continuously 'scans' the stage floor forcing the performers to either jump or dive to the floor as it passes. The event is a multi-layered dense montage whereas in most techno-art performances technology is used mainly to create spectacle. In, PH, it felt as though the viewer was exposed directly to the affects of technological consumerism.

Although the work of Scanner is usually discussed in relation to surveillance, another aspect of his work draws attention to the psychological space of the airwaves. Telephone conversations represent for him a more honest depiction of the world than the outpourings of televisual reality. Long samples of telephone 'normality' "contain more soap opera in the 54 and a bit minutes of Scanner... than there is in a month of Eastenders". (02)

In CRASH the psychopathological nature of everyday technology is explored through a consumer product, the car. Ballard's provocative introductory essay paints a dark picture of the writer's imagination as an "'inner space', that psychological domain (manifest for example, in surrealist painting) where the inner world of the mind and the outer world of reality meet and fuse". In a world "ruled by fictions" the writer's task is to invent the reality. (03)

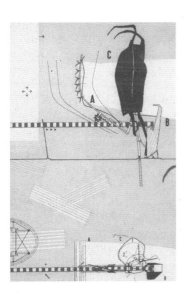

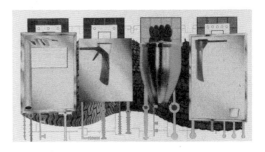

FIG. 4.4
This project by Architektbüro Bolles ·
Wilson (1988) is an architectural exploration
of the poetics of electronically mediated
architectural space, in this case urban space.

FIG. 4.5
Ben Nicholson's book **APPLIANCE HOUSE** (1990) is an architectural
fiction dealing with imaginary psychological narratives derived on one level
from the products on offer in a Sears catalogue and on another from the
imagined structures built by a kleptomaniac.

DESIGNER AS AUTHOR

When we talk about what goes on in a computer, we're talking about an entire complex of relations, assumptions, actions, intentions, design, error, too, as well as the results, and so on. A computer is a device that allows us to put cognitive models into operational form. But cognitive models are fictions, artificial constructs that correspond more or less to what happens in the world.

G. CHAPMAN, **MAKING SENSE** OUT OF NONSENSE

Conventional roles for design include addressing problems set by industry, designing interfaces that seduce the user into cybernetic communication with the corporate cultural values embodied in the emerging environment of digital objects, and finding novel applications for new technologies. But design could also develop new attitudes to electronic technology. To do this, designers could become more like authors, drawing from the narrative space of electronic object misuse and abuse to create alternative contexts of use and need.

Design could explore the fluid interface between "cognitive models [as] fictions, artificial constructs" and new electronic technologies. Designers could create new critical artefacts that help consumers, as protagonists rather than users, to navigate through the "communications landscape" we share with "the spectres of sinister technologies and the dreams that money can buy". (04)

To explore this new role for designers, it is necessary to turn first to architecture. Although the relation of narrative to space has been thoroughly theorised by architects such as Tschumi, and explored through design proposals by Nigel Coates and Narrative Architecture Today (NATO), (05) few architects have considered narrative space within the context of an electronic consumer-driven society and even fewer in a way that specifically addresses experiences centred on electronic products. Michael Sorkin comes close on an urban scale, and Catrina Beevor evokes consumer electronics as a potentially liberating force for narrative architecture. (06) But only a handful of projects have developed actual architectural proposals, usually through the ploy of designing 'a house for...'.

FIG. 4.6
Toyo Ito's DWELLING FOR TOKYO NOMAD WOMAN (1985)
is an architectural response to consumerism which sets his work
apart from many other architects, including those who address
technological issues.

One such is an early exploration of the poetics of electronically mediated architectural space, in this case urban space, by Architekturbüro Bolles + Wilson. This competition entry uses architectural function to respond lyrically to electronic media through the design of a retreat from the "electronic glare" of Tokyo, an electronic shadow **(FIG. 4.4)**:

Encompassed in the concepts of 'electronic glare' and 'electronic shadows' is Ninja Architecture. Devised by Architekturbüro Bolles + Wilson Ninja Architecture describes the function and meaning of projects developed in response to an electronically-dependent consumer society... an architecture in search of method and meaning.

D. DOLLENS IN OFF RAMP

The APPLIANCE HOUSE, a project in the form of a book by Ben Nicholson, is another architectural fiction dealing with imaginary psychological narratives, derived on one level in this case from the products in a Sears catalogue and on another from the imagined structures built by a kleptomaniac **(FIG. 4.5)**. The development of the project through collage means that the reality of a built project is unlikely to match the richness of the book and, when we do see glimpses of realised pieces of furniture, only the figurative elements make the transition from book to object.

Toyo Ito's interest in an architectural response to consumerism sets his work apart from most other architects, including those who address technological issues. He deals with fiction, packaging and the private/public experiences of the city. For example, his proposal for a DWELLING FOR TOKYO NOMAD WOMAN **(FIG. 4.6)** is very different from the 1966-68 visions of Mike Webb's CUSHICLE and SUITALOON. While Webb's nomads carry all their belongings with them, Ito's office girl lives in the city, and her home is merely a floating canister for housing the most basic activities, not belongings. Her life revolves around shopping, and her consumerism generates an architecture that offers a fresh alternative to obsessions with the imagery of consumption. In this architectural fiction, conveyed through photographs she drinks tea, reads magazines and applies make-up — hardly utopia. Through projects like these, architects explore the psychological and behavioural dimensions of consumer culture rather than the technical, formal or structural possibilities of consumer technologies.

FIG. 4.7
The **TRUTH PHONE** by the Counter Spy Shop, is an
example of an object embodying a pathological model
of behaviour. It combines a voice-stress-analyser and
telephone, allowing the user to tell whether the person
at the other end is lying.

PRODUCT AS ROLE MODEL

Examples of how design responds to the psychological and behavioural dimensions of electronics can be found at the edges of anonymous design. Obscure marketing and novel technical possibilities lead unintentionally to objects which, although sometimes gimmicky, offer unusual narrative possibilities. A remote-control watch, addressing an unlikely psychological obsession, speaks of a sad need to control the plethora of domestic gadgetry in not only one's own home but also those of others.

The TRUTH PHONE **(FIG. 4.7)**, a relatively straightforward example of an object embodying a pathological model of behaviour, paranoid suspicion, combines voice-stress-analyser and telephone, allowing the user to tell whether the person at the other end is lying. Although it resembles the absurdist gadgets of Garner, its functional restraint and sober appearance help suspend one's disbelief, something not achieved by many gadgets. The TRUTH PHONE illustrates how an electronic product can transform the perception (and conception) of the user as a protagonist, in this case by embodying unusual psychological needs and desires in pathological objects. When one imagines using this object to talk to lovers or family members, its critical function becomes clear. By imagining living with it the owner explores boundaries between himself or herself and the paranoid user-model embodied in the product.

The TRUTH PHONE is a 'role model' in the sense meant by George Herbert Mead. (07) Through the conceptual model of behaviour embodied in its functionality and operation it allows the user to participate in situations that encourage critical reflection on the socialising effect of our encounters with everyday electronic products. It does this not didactically but in a more ambiguous and indirect way. This and similar electronic objects, generate a conceptual space where interactivity can challenge and enlarge the scheme through which we interpret our experiences of using everyday electronic objects and the social experiences they mediate.

NOTES

(01)

"An urban legend appears mysteriously and spreads spontaneously in varying forms. Contains elements of humour or horror (the horror often 'punishes' someone who flouts society's conventions). Makes good storytelling. Does NOT have to be false, although most are. ULs often have a basis in fact, but it's their life after-the-fact (particularly in reference to the second and third points) that gives them particular interest." (From FAQ alt.urban legends)

(02)

King, TRANCE EUROPE EXPRESS 2, pp. 134-38.

(03)

Ballard, CRASH, p. 5.

(04)

ibid, p. 5

(05)

"Its pursuance of current lifestyle as the sustaining parallel to the design of cities forms the basis of its spirit and optimism." (opening comment on each issue of NATO)

(06)

"So technology might be the passport back to a world where concepts and images are more important than actuality. In the manner of reclusants and hermits of yore, by his handling of soft and hardware, modern man can be a dreamer in meta-mataphysical [sic] space. If however he is unready to discard the reality of this earth in favour of a manufactured romance, at least his technologically induced, media fed sensibilities can be better married to the outside world. If he knows fictional distortions in time and in space, it becomes easier to settle in real places that are learning to use an apposite technique of expressiveness." (Beevor, NATO, p. 6.)

(07)

"Unfortunately, Mead's original meaning of the term 'role model' has become narrowed, so that now social scientists tend to emphasize the behavioral patterns of an actual person as constituting a "role model", leaving out or omitting the fact that Mead includes 'any object' or 'set of objects' as having this power as well." Csikszentmihalyi & Rochberg-Halton, THE MEANING OF OBJECTS, p. 51.

Considered as an operator acting in relation to the daily environment, the designer's ultimate responsibility can only be to contribute to the production of a habitable world, a world in which human beings not merely survive but also express and expand their cultural and spiritual possibilities. The term habitable, referring to the environment, indicates a complex existential condition that cannot be reduced to its functional component. It is a condition arising from the intersection of a multiplicity of questions rooted in the anthropological and social nature of the human race.

E. MANZINI IN DESIGN ISSUES 9 (1)

ibute to the production of a habitable world" design needs to be transformed, expanding its scope to include speculation on how best to provide the conditions for inhabitation. It must not just visualise a 'better world' but arouse in the public the desire for one. Design approaches are needed that focus on the interaction between the portrayed reality of alternative scenarios, which so often appear didactic or utopian, and the everyday reality in which they are encountered.

FICTION

Many issues touched on here, such as art's relation to everyday life, and the need for art to resist easy assimilation, overlap with those already addressed by the Frankfurt School and others in relation to disciplines such as music (Adorno), painting (Marcuse), art (Benjamin) and drama (Brecht). The similarities between these issues and those addressed by Marxist approaches to aesthetics does not imply an identification with Marxism but is the result of seeing design as having value outside the marketplace – an alternative to fine art.

of design can only exist outside a commercial context and, indeed operates as a critique of it. It is a form of 'conceptual design'— meaning not the conceptual stage of a design project, but a product intended to challenge preconceptions about how electronic products shape our lives. This chapter discusses how such design thinking might re-enter everyday life in ways that maintain the design proposal's critical integrity and effectiveness while facing criticism of escapism, utopianism, or fantasy. The challenge is to blur the boundaries between the real and the fictional, so that the visionary becomes more real and the real is seen as just one limited possibility, a product of ideology maintained through the uncritical design of a surfeit of consumer goods.

that this form of conceptual design need not conform to the conventions that shape the design process in relation to the marketplace does not mean it has to be utopian. It can use its independent position to provide conditions that encourage more reflective and challenging design ideas than are possible in commercial design. But if it is to avoid accusations of escapism this design thinking must also develop strategies for linking itself to everyday life that complement those of the marketplace. REAL-FICTION focuses on the problem of 'crossing over', and discusses how conceptual design intentions and formats of work, differ from those of commercial design, and require different contexts in which the design thinking can be encountered by the public. It is concerned with representation and contexts of presentation for ideas about everyday life in the form of

THE DESIGN OBJECT AS PROTOTYPE

There is a danger that if design is not oriented to the marketplace it is seen as invalid, irrelevant or self-indulgent, especially if displayed in a gallery. But what if the gallery were viewed as a test-site for designs unlikely to enter everyday life? What would be the most effective format for a design object designed to be shown in a gallery?

The most obvious would be fully working prototypes that can be 'tested out' on the public in the gallery and, if the reaction is good, later mass-produced. But fully working prototypes displayed in galleries rarely challenge viewer's assumptions about the role of products in their lives. For example, many visitors to my contribution to the MONITOR AS MATERIAL exhibition at the RCA in 1996 **(FIG. 5.1)** said they found the work interesting as spectacle, but had missed concerns with the more fictive, social and aesthetic aspects that linked it to everyday life, even if only conceptual. Its strangeness and apparent interactivity emphasised the here-and-now. The gallery became a 'bracketed space', an abstract setting, disconnecting the experience of engaging with the work from everyday life. Displaying a fully working prototype in a gallery context invites people to marvel at the ingenuity of the designer, and the fact it works, but overlook the challenge to the status quo its insertion into everyday life might bring about. (01) Following this route, the gallery becomes a 'freak show' of objects of wonder and amusement. The electronic objects of Weil, re-interpretations of existing products such as radios, digital clocks and calculators, focus on the conceptual relationship between the person and the electronic object. Displayed in the gallery as one-offs, as objets d'art, they achieve little. But if the prototypes are batch-produced (which Weil's objects were), the gallery becomes a 'showroom', allowing them to enter everyday life through the marketplace: a specialist shop selling state-of-the art material culture, trading in the shock-of-the-new re-interpretations of familiar objects.

THE DESIGN OBJECT AS INSTALLATION

For the designer who regards the electronic object as an embodiment of potential patterns of behaviour and ideology, careful consideration of the relationship between the gallery and the conceptual design object is essential if the object is to connect with everyday life.

FIG. 5.1
In my contribution to the **MONITOR AS MATERIAL** exhibition at the RCA in
1996, many visitors to the space found the work itself interesting as spectacle,
but concerns with more fictive, social and aesthetic aspects that linked the work
to everyday life, even if only through the imaginary, were lost.

Electronic objects that use the gallery to demonstrate their interactive aesthetic or experiential aspects can be subsumed by kinetic art culture whose focus is on the here-and-now and providing an escape from everyday life. An installation by Fiona Raby for Electra 96 highlights this problem. The installation was intended to be a test-site for a design proposal linking two locations by open telephone lines. Ultrasonic sensors registered approaching people and allowed sounds from the other location to filter through, distorted at first they cleared as the person moved closer to the installation so that spoken communication could take place. As a design proposal it would be experienced by a building's inhabitants over several years and the aesthetic experience would have to be very subtle. As an exhibit in an electronic event the installation was expected to provide immediate feedback in an entertaining way. It might have been better to exhibit a film which used non-working props to explore how the proposal might impact over time on the day-to-day experiences of fictional users.

One of James Turrell's projects, PERCEPTUAL CELLS, offers an interesting solution to this problem. Once inside a booth-like structure, a bit like a telephone box in a gallery, visitors are presented with controls to vary the colour of light in a hemisphere surrounding their heads. Humorous and quirky, it invites comparison with street furniture and public utilities, and their association with mass-consumption, state ownership and industrial production. The visitor imagines, perhaps, using one of these machines on the street, so a strong link with the world outside is established. It successfully combines the best qualities of prototypes and installations: it can be used in the gallery rather than just contemplated, and at the same time establishes links with life beyond the gallery.

The ALIEN STAFF by Wodiczko demonstrates another approach — intervention. Wodiczko's project shows how industrial design, by imagining new functions and configuring them as usable prototypes, can function critically outside the gallery. Wodiczko has deployed teams of 'aliens' in various cities armed with his ALIEN STAFF and studied the resulting interactions between them and the public **(FIG.5.2)**.

Such objects, using simple electronic technologies and emphasising invention and social and cultural content, are rare examples of how product design and the electronic object can fuse into design as criticism. The prototype draws attention to the boundaries of normal

FIG. 5.2
Krzysztof Wodickzko's **ALIEN STAFF** (1992) – houses a small
LCD television. The small size of the display, its position at eye level
and its proximity to the alien's face are all important.

behaviour and thought by intervening in everyday social situations outside the gallery. That they are deemed problematic by the design world draws attention to other boundaries of categories of practice and ideas:

Asked how the design world has responded to his various Homeless Vehicle [sic] Wodiczko throws back his head and laughs at the pretensions of the so-called 'designer decade'... "The minute you present a proposal, people think you must be offering a grand vision for a better future". They can't see a thing like the Homeless Vehicle or the Poliscar as the 'concretisation' of a present problem, a makeshift transitional device, or an aesthetic experiment. Instead, "they think it must be designed for mass production, and instantly imagine 100,000 Poliscars taking over the cities".

P. WRIGHT, THE POLISCAR

THE DESIGN OBJECT AS MODEL

What is the potential of non-working design models as opposed to prototypes? The preoccupation with product semantics, that dominated design in relation to electronic objects for most of the 1980s, focused attention on the object itself, particularly its visual meaning. The concept model functioned as a didactic design object, it was not something to challenge the way we lived our lives, but a meta-design challenging only design itself.

The non-working model is the conventional physical representation of conceptual design proposals: naturalistic, non-working mock-ups simulating the appearance of a mass-produced object. Yet this freedom from technical functionality could be better used. If the design model was viewed as a medium in its own right, it could exploit its non-working status to address issues beyond the scope of the technically functional prototype. But to achieve this it needs to be considered as a model in the same sense as a mathematical or cognitive model. This enlarged view of the model is already accepted in architecture and fine art:

The space of the model lies on the border between representation and actuality. Like the frame of a painting, it demarcates a limit between the work and what lies beyond. And like the frame, the model is neither wholly inside nor wholly outside, neither pure representation nor transcendent object. It claims a certain autonomous objecthood, yet this condition is always

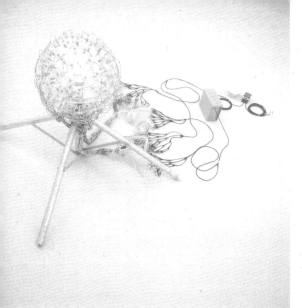

FIG. 5.3
Gregory Green's **NUCLEAR DEVICE #2**
15 KILOTONS, PLUTONIUM 239 (1995) is a model,
a technological object that looks as though it works but does not.
Although it could be made to work, its interest stems from the
fact that the knowledge embodied in these objects is widely
available and very destructive.

incomplete. The model is always a model of. The desire of the model is to act as a simulacrum
of another object, as a surrogate which allows for imaginative occupation.

C. HUBERT, THE RUINS OF REPRESENTATION

In the art world, a huge range of conceptual roles for the model has been explored. Particularly
relevant, because it comes close to that of product designers, is the work of Gregory
Green who builds models of bombs **(FIG. 5.3)**, technological objects that look as though
they work but do not. Although they could be made to work, their interest stems from the
fact that the knowledge embodied in them is widely available and very destructive. The
integration of the 'bombs' as booby traps into familiar objects like suitcases links them to
the world outside the gallery. Their technical uselessness becomes part of their value,
shifting attention to their role as conceptual machines that engage the imagination and
draw the viewer into a reflective and critical space.

These devices look similar to K/K Research and Development's ANALOGS assembled from found
machine parts. But these only work in relation to a narrative, usually social and political,
in an accompanying text. They engage the viewer but are not powerful in themselves.

CRIB-BATIC another project by K/K Research and Development, in collaboration with Scholz, is a
model for a push-chair, an existing object type. We know these objects exist in everyday
life, how they are used, and by whom. So we can imagine what it would mean for their
proposal to enter everyday life. It is not necessary to see the CRIB-BATIC 'working' for it
to be effective, but rather to sense how it might cross over into everyday life. An
imaginative alternative, it is 'fact' in that it could be built, but fiction in that it is unlikely
to be built. This fictiveness enables it to function critically, by highlighting the boundaries
that limit everyday experience. It celebrates the complex ambiguity of the object, as both
part and not part of the society from which it emerged. It has not acceded to the demands
of 'miserable reality' but remains defiantly conceptual.

From a product design point of view these models lack industrial realism; they look like craft
objects, hand-made and probably one-off. But an expanded view of the conceptual design
model might regard it as embodying the essence of a design idea, a 'genotype'[02] rather
than prototype, constructed from the materials at hand. If taken up for mass manufacture

its construction and structure would undoubtedly change. The object's 'content' or 'genes' are important, not its appearance. In the context of design, the conceptual model as genotype rather than prototype could allow it to function more abstractly by deflecting attention from an aesthetics of construction to an aesthetics of use. The genotype depends on the view that a design idea can transcend its material and structural reality and function critically, in relation to social systems for example, rather than visual culture. Andrea Branzi suggests this as a possible role for craft in late-twentieth century industrialised production.(03) Experimental furniture such as Studio Alchymia's 1980 BAUHAUS 2 range **(FIG. 5.4)** do not simulate how they would be if mass-produced, but take a form appropriate for exhibition and consumption as one- or two-offs. Rather than an autonomous form of design, the craft object is seen as one stage in developing a design idea that might eventually be mass-produced.

Michele De Lucchi presented design studies for small domestic electric appliances **(FIG. 5.5)** at the 1979 Milan Triennale. They echoed a contemporary concern to challenge prevailing images of domestic technology. They are interesting because they do not mimic reality; they are clearly representations, 'models' comfortable with their unreality. They are things in themselves rather than shadows of yet to be realised products. They offer real experiences of ideas rather than a unreal experiences of unrealised products, and accept that these ideas will be consumed through books and exhibitions not in the marketplace.

THE DESIGN OBJECT AS PROP

By abandoning the technical realism of the prototype and the visual realism of the traditional industrial design model, conceptual models in combination with other media, can refer to broader contexts of use and inhabitation. For instance, by using conceptual models as film props the viewer can be drawn into the conceptual space of the object in *use* rather than an appreciation of the thing in itself.

Branzi suggests the age of the "Historical Avant-Garde" is ending. Large corporations work with small experimental design centres to develop new scenarios within which the corporations develop new products. He calls this a period of "Permanent Avant-Garde", the aim of which is:

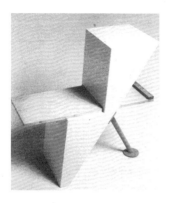

FIG. 5.4

Andrea Branzi's. **GINGER** (1980) for Studio Alchymia
does not simulate how it would be if mass-produced,
but takes a form appropriate for exhibition and
consumption as one- or two-offs. The craft object is
seen as a stage in the development of an idea that
might eventually be mass-produced.

*To restructure the market, to develop a new ecology of the
natural and artificial environment, and to create islands of
meaning that define consumption not as a category of the
emphemeral and provisional, but as a solid culture for a democratic and reformed society, one
in which a new generation of tools will be able to liberate people from uninspiring work,
encouraging mass creativity and individual freedom.*

A. BRANZI, CRISIS IN GROWTH: INDUSTRIAL DESIGN AND THE CREATIVE TRADITION

A key tool in this process is the scenario, both to generate design ideas and communicate the results. Large corporations employ scenarios of use to anticipate how people will interact with the complex environments of which technological products are a part. Usually scenarios have a conservative role, predicting patterns of behaviour in relation to technological developments. They draw from what we already know about people, and so weave new ideas into existing realities. These scenarios extend pre-existent reality into the future and so reinforce the status quo rather than challenging it. For example, WORKSHOP by Philips in collaboration with Olivetti, explores the new office landscape to formulate a new vision of the workplace and propose new tools to support it. But the way it was conceived only reinforced stereotypes of the future office. COMMUNICATOR **(FIG. 5.6)** for 'anywhere anytime' multi-media computing, the GROUP TOOL **(FIG. 5.7)** which encourages office workers to mix fundamental tasks such as photocopying and faxing with socialising, and the fetishistic arrangement of tools that can be interconnected to meet specific functional requirements, propose no innovative vision of changing patterns of work. Corporations need to ensure a continued need for physical products in a world where many products are being replaced by software (for example, phone and fax software for computers). But as a tool for presenting design ideas, the scenario is very powerful. It can draw the viewer into a narrative that goes beyond the object to reveal more complex issues.

Manzini argues that, although design can neither change the world nor create lifestyles that enforce patterns of behaviour onto society, the designer is not simply a problem solver but an intellectual able to link "the possible with the hoped-for in visible form".(04) Manzini's emphasis is less on interaction with discrete objects than on systems of objects. He suggests designers as independent agents use their imaginative skills to propagandise socially and politically desirable situations. In Manzini's view part of the designer's role is democratically to discover what is 'desirable' rather than imposing their own or another minority vision onto society. But Manzini's approach, although critical in that it rejects prevailing conditions and proposes an alternative, runs the risk of being either too didactic or utopian.

FIG. 5.5
Michele De Lucchi's appliances for the 1979 Milan Triennale
do not mimic reality; they are clearly representations;
'models' comfortable with their unreality.

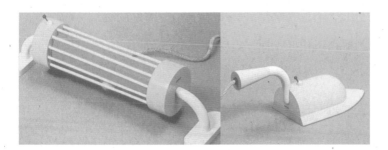

The sci-fi genre offers a third possibility. Susani, noting how what was once called 'concept design' has now become the design of entire scenarios of objects, refers to Apple's 1987 KNOWLEDGE NAVIGATOR project as probably the first use of video narration to present a 'cultural project'. Susani claims it was neither a promotional tool, nor simply a projection of technological evolution, but a study of how we could coexist with new technological artefacts. He suggests that Wim Wenders' film UNTIL THE END OF THE WORLD is a more stimulating and useful project for a 'telephone scenario' than many mainstream design projects for telephones of the future. The use of scenarios in UNTIL THE END OF THE WORLD comes close to being critical because it achieves a degree of estrangement through the behaviour of fictional characters who do not have to conform to existing personality types, occupations or motivations.

But this approach falls foul of a central contradiction of radical work, as Adorno demonstrated in his contrasting of modern classical music and popular jazz. As a mainstream film has to be immediately graspable by a broad audience, the fact of achieving this diminishes its critical potential. Transformations of consciousness are more likely through struggling to understand ideas: simplification dilutes the power to challenge established values:

According to Marcuse, the strength of art lies in its Otherness, its incapacity for ready assimilation. If art comes too close to reality, if it strives too hard to be comprehensible, accessible across all boundaries, it then runs the risk of becoming mundane. And if this occurs, its function as negation to the existing world is abandoned. To be effective art must exert the capacity for estrangement... it must dislocate the viewer, reader, audience, by its refusal and inability to become part of the reality principle.

C. BECKER, HERBERT MARCUSE AND THE SUBVERSIVE POTENTIAL OF ART

If the conceptual design object is to be used as a prop in a scenario that works in a critical, transformative way, other possibilities must be developed. Although a critical approach might alienate some, it might also more effectively transform the consciousness of those whom it does engage. The task is to embody content in an aesthetically challenging form that would "push the viewer towards a more complex, emotional, or revolutionary understanding of the problems posed by the work".

FIG. 5.8

Cindy Sherman's UNTITLED
FILM STILLS series (1978) shift
the viewer's imagination towards the
fictional possibilities of the portrayed
moment. The furnishings and
incidental objects in these photographs
encourage an allegorical reading that
further engages the viewer.

FIGS. 5.6-7

Philips/Olivetti's COMMUNICATOR and GROUP TOOL (1994)
set out to explore the new office landscape, to formulate a new
vision of the workplace, and propose new tools to support it.

Some artists and sculptors have achieved this in films they have made about their work (Philippe Ramette, for instance, or Rebecca Horn), and filmmakers such as David Lynch have developed strategies for applying this to television (for instance, Lynch's TWIN PEAKS). But there are few examples (one being, Atom Egoyan's FAMILY VIEWING) where electronic products play a significant role.

Cindy Sherman's photographs from her UNTITLED FILM STILLS series **(FIG. 5.8)** portray banal moments of apparently little significance. As the viewer tries to imagine what happened before and after they are drawn into speculation on the psychology of the protagonists and their state of mind. These photos show the surprising power of stills, compared to video or film, to engage the viewer. They shift the viewer's imagination towards the fictional possibilities of the portrayed moment. The furnishings and incidental objects in these photographs encourage an allegorical reading that further engages the viewer. Most of the images look as though they were taken in the 1950s or 1960s which adds to the distance they create.

The work of Garner also uses a sense of the recent past to engage audiences. His two books, consist of photographs **(FIG. 5.9)** of conceptual design objects, and of scenes reflecting the strange psychological social narratives that arise from interaction with and through the objects portrayed. The books could be seen as a critique of consumer society, but their dependence on comic absurdity distracts attention from any serious criticisms that might be read into the project. This ironic approach offers no constructive suggestions. In comparison, Ito's DWELLING FOR A TOKYO NOMAD WOMAN, an architectural fiction conveyed through photographs, portrays a system behaviour and consumption used to make familiar but exaggerated consumer values real and concrete, values that are neither futuristic nor utopian, but uncomfortably close to our own. The nomad woman's only furniture is designed to support intelligence gathering on new trends, eating snack food, and styling one's image. Ito's photographs conjure up an 'elsewhere', familiar but different. Rather than offering another option, or parodying what exists, they suggest that the way things are is not the only possibility.

Although far more nostalgic and romantic, the images produced by Ramette, of himself using his inventions, work in a similar way **(FIG. 5.10)**. The style of his images is deliberately

FIG. 5.9
Philip Garner's **UTOPIA OR BUST** (1985) consist of photographs of conceptual design objects, and of scenes that present unusual narratives that arise from interaction with and through the objects portrayed.

straightforward, and the use of his devices, which usually resemble nineteenth-century scientific instruments are easy to understand. The viewer wonders at the strangeness of Ramette's behaviour, trying to imagine why somebody would behave like this, what pleasure they have, and what prevents such objects being widely disseminated and the values they embody gaining general acceptance.

In Horn's films, DER EINTANZER (1978) and LA FERDINANDA: SONATA FOR A MEDICI VILLA (1981), her sculptures appear in the background of several scenes **(FIG. 5.11)**. They are never explained, but the viewer is drawn into a strange world which objects such as these seem to inhabit nonchalantly. The films seem set in the present, but the integration of such strange objects into everyday settings implies a completely different set of cultural and aesthetic values highlighted by their familiar settings. This technique is reminiscent of Brechtian alienation, in this case drawing our attention to the role of objects in defining and realising everyday space and rituals. Horn's films are neither didactic nor utopian, nor are they parodies. They seem closer to heterotopias. They portray situations different from our own where enchanted objects have a place in daily life and a different 'sense' prevails, a sense interwoven with our own rather than completely alternative or nonsensical. Norman Daly's THE LOST CIVILISATION OF LLHUROS is an exhibition of artefacts from a fictional culture, each of which is accompanied by a caption explaining what is supposedly known about it. The exhibition blurs the boundaries between imaginary spaces and the here-and-now of the gallery. It is as though a film has re-entered everyday life through its props. It invites the visitor to speculate, as an anthropologist of material culture might, on how values come to be embodied in artefacts.

CONCEPTUAL CONSUMERISM

For Marcuse, art is a location – a designated imaginative space where freedom is experienced. At times, it is a physical entity, a site – a painting on the wall, an installation on the floor, an event chiselled in space and/or time, a performance, a dance, a video, a film. But it is also a psychic location – a place in the mind where one allows a recombination of experiences, a suspension of the rules that govern daily life, a denial of gravity. It 'challenges the monopoly of the established reality' by creating 'fictitious worlds' in which one can see mirrored that

range of human emotion and experience that does not find an outlet in the present reality. In this sense the fabricated world becomes 'more real than reality itself'. Art presents the possibility of a fulfilment, which only a transformed society could offer.

C. BECKER, HERBERT MARCUSE AND THE SUBVERSIVE POTENTIAL OF ART

This chapter has discussed where this space might lie in relation to the electronic as conceptual design object, and how we might encounter it. As a route for developing critical electronic objects within a design context, it has rejected the prototype in favour of combining non-working models with film, video or photography to establish scenarios that are neither didactic nor utopian but heterotopian. Were the props from a scenario physically displayed with the film, video or photograph, more subtle interactions might develop between the space of the here-and-now, where the viewer is, and the fictional space portrayed in the image. The physical presence of the artefacts encourages additional interplay between reality and fiction, between what is and what might be. By themselves the artefacts would be mentally assimilated into known patterns of behaviour, 'explained away'. But shown as part of an alien culture with different aesthetic values and a different 'sense', they require viewers to accommodate the unusual role of the artefacts in an everyday life like their own.

The space in which the artefacts are shown becomes a 'showroom' rather than a gallery, encouraging a form of conceptual consumerism via critical 'advertisements' and 'products'. New ideas are tried out in the imagination of visitors, who are encouraged to draw on their already well-developed skills as window-shopper and high-street showroom-frequenter. The designer becomes an applied conceptual artist, socialising art practice by moving it into a larger and more accessible context while retaining its potential to provoke people to reflect on the way electronic products shape their experience of everyday life.

N●TES

(01)

Although it is not an electronic product, when Krzysztof Wodiczko's POLISCAR was exhibited at the Josh Baer Gallery, he was criticised for "'revelling in the mechanics' of his warlike contraption when he might be more usefully have been working out how it could 'transform current social realities'". Yet that is exactly what the artist was concerned with. P. Wright, "The Poliscar: Not a Tank but a War Machine for people without Apartments".

(02)

I use the term 'genotype' as an alternative to 'prototype' to shift importance away from whether or not a conceptual design technically works, to the ideas it represents.

(03)

A. Branzi, THE HOT HOUSE, p. 141.

(04)

Manzini, "Design, Environment and Social Quality: From 'Existenzminimum' to 'Quality Maximum'" in DESIGN ISSUES, 1994, vol. 10, no. 1, pp. 37-44.

It might seem strange to write about radio, [01] a long-established medium, when discussion today centres on cyberspace, virtual reality, networks, smart materials and other electronic technologies. But radio, meaning part of the electromagnetic spectrum, **(FIG. 6.1)** is fundamental to electronics. Objects 'dematerialise' not only into sofware in response to miniaturisation and replacement by services, but literally dematerialise into radiation. All electronic products are hybrids of radiation and matter. This chapter does not discuss making the invisible visible, or visualising radio, but explores the links between the material and immaterial that lead to new aesthetic possibilities for life in an electromagnetic environment. Whereas cyberspace is a metaphor that spatialises what happens in computers distributed around the world, radio space is actual and physical, even though our senses detect only a tiny part of it.

≈ 06
HERTZIAN
SPACE

It is just over 100 years since electricity generation started, 70 since radio transmissions began, and 50 since radar and telecommunications entered our environment. The twentieth century has seen space evolve into a complex soup of electromagnetic radiation. The extra-sensory parts of the electromagnetic spectrum form more and more of our artefactual environment, yet designers direct little attention towards the possible sensual and poetic experience of this industrially-produced new materiality. [02] In design, immateriality is often referred to through visual motifs **(FIG. 6.2)**, usually in relation to product semantics and representation, but is rarely dealt with directly as a physical phenomenon.

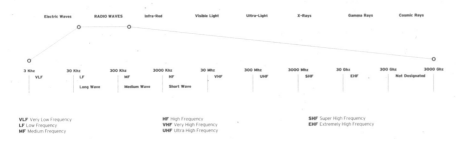

TUNEABLE REALITY

The extra-sensory nature of electromagnetic radiation often leads to its treatment as something conceptual — which easily becomes confused with the notional, although of course it is physical and exists in space. The conflict between the conceptual and the perceptual aspects of hertzian space is an appropriate vehicle for investigating the boundaries between the imaginary and the actual.

Lee and Dawes exploit this ambiguity in IN THE ETHER, a combination of film, theremin music and performance, which takes the audience into a nostalgic and surreal realm between fiction and actuality:

They like it up high, radio waves. If you attach a long piece of wire to a long pole and put it up high you can hear them. I imagine that they spend all their time racing each other around and around the earth. I don't expect they come down very often. Except if they are curious.

LEE & DAWES, IN THE ETHER

Their film, made with Frances Boyle, taps into vaguely paranormal myths of radio folklore concerned with the mysteries of magnetism, and offers what can be described as a 'psycho-hertzian' reading of everyday life.

AERIAL PARIS and AERO-LIVING LABORATORIES by Lebbeus Woods, a re-siting of architecture in electromagnetism, exploit this ambiguity less successfully. Electromagnetism becomes a field which "binds building to the sky instead of the earth".(03) Although one of the few architectural propositions centred on the electromagnetic aspects of space, this "architecture suspended in an invisible matrix of air and charge" is a form of science fiction. Its grand speculations and escapist logic cannot match the gently provocative poetry of Lee and Dawes. It is difficult to see exactly what Woods' two projects gain through their association with electromagnetism in terms of architecture or new models of living .

Another architect, Laura Kurgan, responds more directly to inhabiting a ubiquitous electrosphere projected onto earth by a network of satellites.(04) Using what Virilio calls "the little everyday object [that] probably constitutes the event of the decade as far as globalisation of location goes", the GPS (Global Positioning System) navigator, she rigorously maps her

FIG. 6.2
In design, immateriality is often referred to through visual motifs usually
in relation to product semantics and representation, but is rarely dealt with
directly as a physical phenomenon. This telephone is meant to reflect the
"imagery of the microwave communications systems to which it is attached".

explorations of this space. The GPS, which uses military satellites to plot the position of a sensor anywhere on the planet, is currently not very accurate, partly because for security reasons the military do not want civilians to have access to such accuracy and partly because the signal is distorted through reflections near the sensor, for instance by buildings. Kurgan uses the GPS to map a space somewhere between the physical, digital and conceptual. She stands in a gallery stationary for ten minutes recording 311 position records, plots the results on a map of the gallery and its surroundings and compares them with a more accurate computer corrected version. [05]

The artist Ingo Günther's sitecasting describes a situation where the television signal does not travel to where you are; the reverse happens. You have to go where it is; you have to hunt for it. He imagines a city of tiny television transmitters broadcasting the forgotten pasts of buildings, places and streets. The city becomes a tuneable urban environment. Different time periods could be arranged as different channels into which the participant could tune. To do this it is necessary to design aerials more directional than usual, allowing different signals to be spatially separate in one location. The resultant antennas resemble the sculptures of Klaus vom Bruch who as early as 1984 exhibited constructivist-inspired aerials for broadcasting video signals between elements of his installations. For Günther's video installation, EXHIBITION ON AIR, the visitor enters the P3 art and environment gallery in Tokyo and wanders about holding a combination of aerial and LCD television, receiving broadcasts from other antennas sited around the building.

A different kind of narrative space is explored by Scanner, who uses a wide-band radio scanner to tune into cellular telephone conversations, combining them on CDs to create ambient and often poignant sound images of the psychological and social poetry of everyday radio space.

These urban analyses of the militarisation of the spectrum, tuneable urban narratives and audio snapshots of telephone 'normality' contrast with the New Age mysticism of sculptor Michael Heivly who, like Woods, seems to encourage an escape into the fantastic rather than a confrontation with and possible transformation of existing reality. Heivly writes about how microwave form, detectable yet unavailable to the senses, [06] represents simultaneously the idea of the real and the imagined. His work translates landscapes into

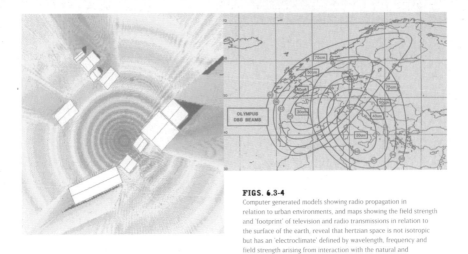

FIGS. 6.3-4
Computer generated models showing radio propagation in
relation to urban environments, and maps showing the field strength
and 'footprint' of television and radio transmissions in relation to
the surface of the earth, reveal that hertzian space is not isotropic
but has an 'electroclimate' defined by wavelength, frequency and
field strength arising from interaction with the natural and
artificial landscape.

musical sound compositions transmitted as microwaves into deep space. The microwave
energy becomes a cone-shaped sculptural form that moves through space at the speed of
light and retains its form for millennia. All his work aims to create an environment that
confronts participants with the known and the unknown, and requires them to use their
imaginations to construct the piece in deep space.

ELECTROCLIMATES

Computer models showing radio propagation in relation to urban environments **(FIG. 6.3)**, and
maps showing the field strength and 'footprint' of television and radio transmissions in
relation to the surface of the earth **(FIG. 6.4)**, reveal that hertzian space is not isotropic but
has an 'electroclimate' defined by wavelength, frequency and field strength arising from
interaction with the natural and artificial landscape.

The extent of the electrosphere is reflected in the difficulty of finding electromagnetically
unpolluted parts of the globe as sites for intelligence gathering, 'antenna farms', the use
of faraday cages to create 'empty' zero-field spaces for isolating sensitive equipment, and
the realisation that a modern war is won by the side that best exploits the electromagnetic
spectrum by denying the enemy its effective use and protecting friendly electromagnetic
systems against electronic attack:

Before we bipedal apes invented radio receivers, before we even exchanged our gills for lungs,
there was radio. It was in lightning, in hydrogen atoms, in the big bang that propelled our
universe into existence. But as soon as we invented technology that enabled us to listen to the
transmissions of our planet, we saturated the airwaves with our own sounds – garage door
openers, cordless phones, baby monitors, police dispatchers, pagers, and wireless microphones
– jamming the oldest radio station around.

N. STRAUSS, RADIOTEXT(E)

'Whistler hunters', natural radio enthusiasts who search out radio transmissions created by
atmospheric events, map the interface between atmospheric and electromagnetic
climates. They search out natural radio signals, VLF (very low frequency) radio waves or
'sferics' (short for atmospherics: natural radio-frequency emissions in the ionosphere,

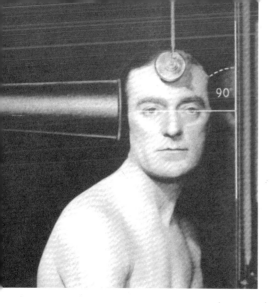

FIG. 6.5
An Ilford manual for x-ray machines contains images
of radiographic actors and props that illustrate the use
of the body as a radio medium.

caused by electromagnetic energy radiated from lightning). These signals — resonant clicks and pops called 'tweaks' and 'bonks' by scientists — occur in the audible range and may be picked up by antennas and amplified for listening. They are best received at night, far from power lines. Occasionally sferics get caught on, and travel long distances along, the magnetic flux lines around the earth, producing 'whistlers', downward gliding signals which may last up to three seconds. Whistler hunters travel far from power lines and electromagnetic pollution, sometimes camping out for days, listening for the elusive sounds of natural radio.

Between 1967 and 1975 the composer Alvin Lucier became interested in these sounds and made performances using pre-recorded whistlers. In 1981 he recorded whistlers and spliced together short samples in chronological order for SFERICS (1988). Other variations were produced for performances during the early 1980s, one of which involved setting up a small array of antennas at a campsite for the public to listen to in real time through battery-powered tape recorders and headsets. But it is doubtful that such artificial events capture the poetry of the whistler hunters' activities. Although the sounds are fleetingly beautiful, out of context they lose much; their beauty is entwined with the effort endured and the symbolic significance of receiving them, which for some is quasi-mystical, for others a defiant gesture against people's careless attitude towards nature.

More successful if less romantic celebrations of the electroclimate of artificial radio have been achieved through radios used as performing instruments by other composers. This began with Cage's IMAGINARY LANDSCAPE NO. 4 for 12 receiving sets: the arbitrary nature of broadcast material must have appealed to the composer of the MUSIC OF CHANGES and 4'33". Later, in KURZWELLEN (Shortwaves) of 1968, Stockhausen used radio sounds to open himself to a "music of the whole earth":

What can be more world-wide... more ego-transcending, more all embracing, more universal and more momentous than the broadcasts which in Kurzwellen take on the guise of musical material?... What happens consists only of what the world is broadcasting now; it issues from the human spirit, is further moulded and continually transformed by the mutual interference to which all emissions are subject; and finally it is brought to a higher unity by our musicians in their performance.

P. GRIFFITHS, MODERN MUSIC

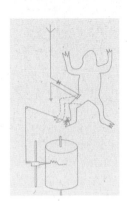

FIG. 6.6

Long before radio energy
was used to carry an acoustic
signal, many ingenious devices
were invented to detect radio
energy. Lefeuvre's 'physiological'
receiver, for instance, uses
the electrical sensitivity of
a frog's leg.

FIG. 6.7

Several of Ito's works attempt to evoke an implied
sensuality. In his TOWER OF WINDS (1986) —
realised in the middle of a neon downtown, in front
of Yokohama station — he wanted the "air itself to
be converted into light".

Whereas these composers celebrated the ubiquity of hertzian space, for Architekturbüro Bolles + Wilson the electronic glare of an invisible ephemeral city of ubiquitous impulses is something to be sheltered from, where comfort is to negate for a moment this network to create a zone of electronic shadows. Their proposal is less about the poetics of revealing the world as it is, and more about charging architectural space with psychological dimensions derived from acknowledging hertzian space.

IMMATERIAL SENSUALITY

We are experiencing a new kind of connection to our artefactual environment. The electronic object is spread over many frequencies of the electromagnetic spectrum, partly visible, partly not. Sense organs function as transducers, converting environmental energy into neural signals. Our sense organs cannot transduce radio waves or other wavelengths outside the narrow bandwidth of visible light (and infra-red energy through the skin as warmth). Electronic objects are disembodied machines with extended invisible skins everywhere. They couple and decouple with our bodies without us knowing. Working on microscopic scales, often pathogenic, many electromagnetic fields interfere with the cellular structure of the body. Paranoia accompanies dealings with such hertzian machines. How do they touch us? Do they merely reflect off our skin, or the surface of our internal organs? In other words, do they merely 'see' us, or can they 'read' us too, extracting personal information about our identity, status, and health?

An operating manual for x-ray machines contains images of radiographic actors and props **(FIG. 6.5)** that view the body as a radio medium. The machines establish views, and support a sort of radio perspective, revealing, concealing and exposing hidden organs and views, and creating a 'radio theatre' of the hidden body. In configuring the body according to an unusual conception of space, these images of people and x-ray machines illustrate an expanded view of space as an electromagnetic medium.

The artist Arthur Elsenaar inspired by photographs of experiments by gentleman scholars in the 1850s, taps into our strong feelings about electricity, its danger and mystery, and its measurability. He uses the microwave field of a radar sensor to create "an aura, or an extension of my skin into spaces, into which people can walk" which causes a 24-volt

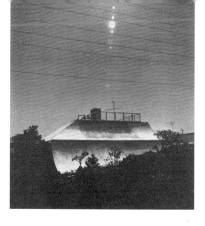

FIG. 6.9
Kazuo Shinohara's **HOUSE UNDER HIGH VOLTAGE LINES** (1981) provides a conceptually eloquent response to the new technological situation brought about by electromagnetic space.

FIG. 6.8
Toyo Ito's **DREAMS ROOM** (1991-92) for the **VISIONS OF JAPAN** exhibition at the Victoria and Albert Museum in London evokes the immaterial sensuality of the new information environment.

pulsed DC current to deliver a variable charge of up to three milliamps to two electrodes attached above the jaw and two to the 'hunch' muscles in the shoulders. He is developing a digital system which will support a wider range of inputs and outputs, for example different responses for people retreating and approaching, and head turns and nods.

Elsenaar's poetic use of fields, a pathological exploration of personal space, is very different from research carried out at MIT into technically interesting but aesthetically mundane applications of electrical fields to inter-personal information exchange. The body is treated literally as a circuit board and the commercial ambitions of the project have eradicated any possibility for poetry although the "electrical whispers of fish" are one inspiration for the project. (07) The potential of the technology is reduced to the most basic level of utility and conceived as a replacement for physical connections between personal databases.

MAKING VISIBLE THE INVISIBLE

Long before radio energy was used to carry an acoustic signal, many ingenious devices were invented to detect radio energy, (08) for example Lefeuvre's 'physiological' receiver **(FIG. 6.6)** that uses the electrical sensitivity of the frog's leg. These objects resemble the early meteorological equipment used to make visible atmospheric phenomena otherwise too subtle for our bodies to sense. Just as the barometer tells us how heavy the sky is, these early radio detectors embody a more poetic understanding of hertzian space by revealing the extent of its presence.

Today a typical urban radio frequency environment is dominated by radio and television broadcast transmissions. Other forms of radiation such as microwave relay links, radio telephones, CB, speed-detecting radar, satellite communications systems, military tracking radar, civilian air traffic control, air route surveillance, and weather radar all combine into what Ito has called "Active Air". (09)

Several of Ito's works evoke this implied sensuality. In his ᴛᴏᴡᴇʀ ᴏꜰ ᴡɪɴᴅs **(FIG. 6.7)** — realised in 1986 in the middle of a neon downtown, in front of Yokohama station — he wanted the air itself to be converted into light. The tower appears to dematerialise at a particular moment, re-appearing in response to ambient noise levels. His ᴅʀᴇᴀᴍs ʀᴏᴏᴍ **(FIG. 6.8)**,

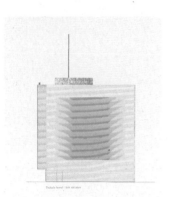

FIG. 6.10
Herzog & De Meuron's **SIGNAL BOX** #4 (1991-94)
is an example of how sensual material responses
to immaterial electromagnetic fields can lead to
new aesthetic possibilities for architecture
situated within hertzian space.

installed at the Victoria and Albert Museum in London in 1991-92, tries to evoke the immaterial sensuality of the new information environment by combining an information-saturated environment of projected imagery with specially commissioned interface objects[10], intended to reinforce at an intimate scale what the environment communicates at the scale of architecture.

Although at first sight the SIGNALS by Takis look like antennas responding to the contents of the air, low-tech precursors of Ito's TOWER OF WINDS, they are in fact non-working symbolic evocations. His work is important not only because it dealt with the poetry of electromagnetism in the 1950s long before others, but also because he developed a language that referred indirectly to the mysterious and metaphysical aspects of electricity and magnetism, in contrast to the more exuberant responses to technology of many of his contemporaries.

Ito's and Takis' pieces visually imply they are translating the invisible into the visible. Another piece of architecture, the HOUSE UNDER HIGH VOLTAGE LINES **(FIG. 6.9)** by Kazuo Shinohara, at the other end of the technological scale from Ito, provides an equally beautiful but more restrained response to the new technological situation brought about by electromagnetic space. The site is beneath high-voltage power lines. Strict regulations determine the safe distance from these lines, and the roof of the house defines this zone for two cables, creating an interface between a possibly pathogenic electromagnetic field and a sculpted interior space. It evokes more disturbing and powerful notions of radio space than the work of either Ito or Takis because it exposes the possible harmfulness of these fields and their existence in everyday life.

The architects Herzog & De Meuron respond to an electromagnetic context in their design for a signal box **(FIG. 6.10)**. The building houses sensitive electronic equipment and needs to be protected from sudden bursts of electromagnetic radiation. It is not clear if this proposal is an alternative or an augmentation of the usual techniques of shielding, but it has resulted in a powerful image of architecture situated within hertzian space. Although relatively low-tech and programmatically mundane, this is another example of how sensual material responses to immaterial electromagnetic fields can lead to new aesthetic possibilities for architecture. It contrasts with more self-conscious rhetorical expressions of electronic culture by architects such as Jean Nouvel, Rem Koolhaas and Bernard Tschumi.

FIG. 6.11

The **F-117** is an object designed to straddle the worlds of electromagnetism and materiality. It is 'tele-dynamic', designed to fly undetected through fields of radar-frequency radiation. These planes refer to fusions of abstract digital, hertzian and atmospheric spaces.

FIGS. 6.12-13

The (Shelf) Loop and Frame Antenna and (Bottle) Dielectric Clad Antenna are examples from DIY books on antenna theory. They generate the kind of pleasure associated with making do and getting by and people's ability to subvert object types and act in new ways on the environment.

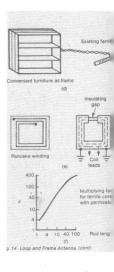

Existing ferrite

Convenient furniture as frame
(d)

Insulating gap

Pancake winding

Coil leads
(e)

400
100
F 40
10
4
1

1 4 10 40 100
(f)
Rod leng

3.14 Loop and Frame Antenna (cont)

THE RADIOGENIC OBJECT

Objects designed to straddle both material and immaterial domains arouse curiosity about the fit between these worlds. Many military aircraft **(FIG. 6.11)** are now 'tele-dynamic', designed to fly undetected through fields of radar-frequency radiation. But tele-dynamic forms are not aerodynamic and to remain airborne their outline needs to be constantly adjusted by a computer. These aircraft fly through fusions of abstract digital, hertzian and atmospheric spaces. If this awareness of hertzian space is to form the basis of an approach to everyday objects, it is not enough simply to present the technical facts. They must be grounded in rich cultural contexts if they are to be more than mere illustrations.

Objects which I call 'radiogenic' function as unwitting interfaces between the abstract space of electromagnetism and the material cultures of everyday life, revealing unexpected points of contact between them. Many of these objects centre on the aerial, a device that links the perceptible material world to the extrasensory world of radiation and energy.

'Aerialness' is a quality of an object considered in relation to the electromagnetic environment. Even the human body is a crude monopole aerial. Although in theory precise laws govern the geometry of aerials, in reality it is a black art, a fusion of the macro world of perception and the imperceptible world of micro-electronics. Embodying the contradictions and limits of scientific thought, an aerial's behaviour can be described but not easily understood because it depends on the dual concept of electromagnetic radiation as wave and particle. As the aerial allows this invisible world to be understood and modelled in terms of material reality, it provides a starting point for a design approach that links the immaterial and the material so as to open up new aesthetic and conceptual possibilities.

Although few artists have explored radiogenic objects, several objects have been created by radio amateurs by enhancing radiogenic qualities in existing environments and artefacts, resulting in objects that provide new perceptions of our hertzian environment. These objects hint at the fertile territory beyond the designer's concern with the semiotics of radio interfaces and the engineer's narrow conception of functionality.

DIY books on antenna theory and practice offer many examples which generate the kind of pleasure which Wentworth notes in MAKING DO AND GETTING BY and people's natural ability

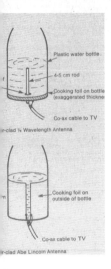

Plastic water bottle
4-5 cm rod
Cooking foil on bottle (exaggerated thickness)
Co-ax cable to TV
Cu-clad ¼ Wavelength Antenna

Cooking foil on outside of bottle
Co-ax cable to TV
Cu-clad Abe Lincoln Antenna

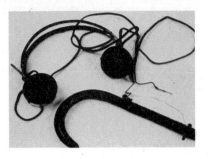

FIG. 6.14
Droz-Georget's BOBBIN CANE was made for listening to the forbidden French transmitter on the Eiffel Tower during World War One.

to subvert object types and act in new ways on the environment. It is a pleasure derived from invention as poetry, loosening the connection between language and things, and challenging the tyranny of language over artefacts. For instance, the use of a specific size of domestic shelving as a core antenna **(FIG. 6.12)** or a milk bottle and tin foil for another **(FIG. 6.13)** reveal unexpected functional connections between physical objects and hertzian space that offer an alternative to representation.

Another example is the BOBBIN CANE **(FIG. 6.14)**, conceived and made by Georges Droz-Georget for listening to the forbidden French transmitter on the Eiffel Tower during World War One. A hook passing through the hole of the ferrule was attached to a low overhead telephone line connected to a shooting range: this became his antenna. The wavelength was selected by moving two sliding rings over the copper thread wound around the shaft, and the receiver was carried in the user's pocket. All these radiogenic objects are part of a hertzian culture that includes diagrams on the use of drain pipes as antennas, and garden layouts **(FIG. 5.15)** which integrate an antenna with vegetables and paths.

WHEN OBJECTS DREAM...

Although when we look at an electronic product we only see what is radiated at the frequency of visible light, all electronic objects are a form of radio. If our eyes could see (tune into) energy of a lower frequency these objects would not only appear different but their boundaries would extend much further into space, interpenetrating other objects considered discrete at the frequency of light. Besides the obvious harmfulness of x-rays and microwaves there is a growing concern over the effect of the radiation leaked by domestic appliances. RADIO AND BEANS, an installation by Patrick Ready, draws attention to the possibly harmful effects of electromagnetic fields generated by domestic appliances. It consists of electrical devices suspended on wooden shelves from the gallery ceiling. Around them hang small paper bags containing fast-growing mung beans in soil, arranged at equal intervals in a three-dimensional grid, and watered three times a day. It was hoped that the beans would exhibit effects from the electrical fields through irregular growth patterns. But as the experiment was not controlled and scientific but ironic, it was never clear how the beans were affected.

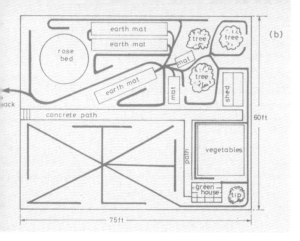

FIG. 6.16

The BAT BAND CONVERTER from EVERYDAY
PRACTICAL ELECTRONICS is a parasitical device that
allows you to "use your AM portable radio and this novel design
to tune-in to the secret world of bats".

FIG. 6.15

Design layout for a garden allowing for the inoffensive
integration of an antenna with vegetables and paths.

In this piece the artist becomes a radio biologist investigating the interaction between radiant energy and biological systems. Science and folklore meet in this strange electrical garden, reminding us of the interconnectedness of nature and technology, something which must be made more visible if we are to find more meaningful ways of inhabiting an environment gradually becoming more radioactive.

The electronic object is often described as 'smart'. But using this term to describe objects with enhanced electronic functionality encourages a bland interpretation of electronic objects.

Smart, after all, is not the same as intelligent, let alone intellectual. Smartness is intelligence that is cost-efficient, planner-responsible, user-friendly, and unerringly obedient to its programmer's designs. None of the qualities, in other words, which we associate with free-thinking intellectuals.

A. ROSS, THE NEW SMARTNESS

Electronic objects are not only 'smart', they 'dream' – in the sense that they leak radiation into the space and objects surrounding them, including our bodies. Despite the images of control and efficiency conveyed through a beige visual language of intelligibility and smartness, electronic objects, it might be imagined, are irrational – or at least allow their thoughts to wander. Thinking of them in terms of dreaminess rather than smartness opens them to more interesting interpretations.

For example, some possibilities for new relationships with these hybrids of radiation and matter are found in pathological products based on paranoia or eccentricity. Many devices designed to transform private situations into public ones depend on the 'leakiness' of electronic objects, tuning into the dreams of radiant objects. The COMPUTER INTERCEPT SYSTEM sold by the Surveillance Technology Group is an example:

Without entering the premises, electromagnetic radiating from unshielded computer screens and ancillary equipment can be intercepted from a remote location. The Computer Intercept System's highly sensitive receiver logs all radiating signals into its 100 channel memory. These emissions are then stabilised, processed and reassembled into clear reproduction of the intercepted data onto its built-in monitor.

SURVEILLANCE TECHNOLOGY GROUP, COVERT AUDIO INTERCEPT

Many buildings are now designed as faraday cages to prevent such eavesdropping, usually invisibly by deploying electromagnetic shielding materials throughout the structure. The same technology protects sensitive equipment in a building from bursts of external radiation. Test-sites, specially designed environments or anechoic chambers, now measure an object's leakiness to predict its effect on other objects.

A more bizarre use of leakiness is seen in the BAT BAND CONVERTER (**FIG. 5.16**), a parasitical device that allows you to "use your a.m. portable radio and this novel design to tune-in to the secret world of bats". The title of the magazine that provides the plans, EVERYDAY PRACTICAL ELECTRONICS, seems at odds with a world where practical skills are turned towards poetic ends, and tuning in to bats, hair, fizzy drinks, crinkly plastic bags and dropping pins is regarded as a sane everyday activity. The device converts the non-electromagnetic ultrasonic signals of the bats into radio signals which are transmitted/leaked to the host radio.

The seemingly illicit information exchange of 'dreamy objects' offers one possible interpretation of the electrosphere. It helps us think of electronic objects in 'hertzian' terms, as interconnected fields rather than discrete things. It acknowledges the problematic conceptual status of electronic objects, arising from their ambiguous identity as hybrids of matter and radiation, functioning at scales and speeds well beyond the range of human perception. If the electronic object has a role in humanising hertzian space it is not as a visualisation or representation of radio but as a catalyst, encouraging the poetic and multi-layered coupling of electromagnetic and material elements to produce new levels of cultural complexity.

NOTES

(01)
Not only radio broadcast frequency but general Radio Frequency: 10 kilohertz – 0.1 terahertz.

(02)
Michele Bertomen's TRANSMISSION TOWERS ON THE LONG ISLAND EXPRESSWAY occasionally mentions electromagnetic space but is primarily a typology of the physical structures that support it. The cultural history of the radio-frequency part of the electromagnetic spectrum and its many uses are well documented in Augaitis and Lander's RADIO RETHINK and Strauss' RADIOTEXT(E). More philosophical and political aspects have been dealt with by Virilio in THE LOST DIMENSION and THE ART OF THE MOTOR, and Manual DeLanda's WAR IN THE AGE OF INTELLIGENT MACHINES.

(03)
M. Sorkin, "Architecture Rising, Lebbeus Woods' Paris Project" in DIDALOS, December 1990.

(04)
L. Kurgan, "You Are Here: Information Drift" in ASSEMBLAGE 25, 1995, pp. 14-43.

(05)
"Download the data to a computer from the receiver, then download from a local base station its (scattered) readings for the same time period (Reference). Correct your readings and reduce the drift (Correction), average the points, and learn where you were – within a few meters (Average). In the computer, the satellites draw the points for you; as the readings become more precise the points grow to fill the screen." Kurgan, op. cit., p. 18.

(06)
M. L. Heivly & M. Reed, "The Space Between the Real and the Imagined: Microwave Sculpture in Deep Space" in LEONARDO, 1992, vol. 25, no. 1, pp. 17-21.

(07)
Some fish use the three-dimensional equivalent of radar to guide themselves through dark waters by positioning themselves in each others' fields, possibly communicating through voltage patterns and rates of pulsing, a sort of electrical whisper. L. Milne and M. Milne, THE SENSES OF ANIMALS AND MEN, pp. 109-20.

(08)
For more on early radio detectors, see V. J. Philips, EARLY RADIO WAVE DETECTORS.

(09)
Although I have not encountered this term in any of Ito's writings he used it often during the design of the DREAMS ROOM for the V&A museum in London.

(10)
Five 'Interactive Terminals' were designed by myself and Fiona Raby for the exhibition.

he manufacture d'Armes de Saint-Etienne] catalogue itself, however - as
ctual existence – is rich in meaning: its exhaustive nomenclatural aims
have the resounding cultural implication that access to objects may be
obtained only via the pages of a catalogue which may be leafed through 'for
the pleasure of it', as one might a great manual, a book of tales, a menu.

J. BAUDRILLARD, THE SYSTEM OF OBJECTS

ion is a commentary on five conceptual design proposals for post-
optimal electronic objects: ELECTROCLIMATES, WHEN OBJECTS DREAM...,
THIEF OF AFFECTIONS, TUNEABLE CITIES, and FARADAY CHAIR.

posal is a material tale, a process of investigation. They are 'value-fictions': they try to maintain a degree of
echnological realism while exploring values different from those current. Their subject is the role of electronic
objects in the aesthetic inhabitation of a rapidly dematerialising, ubiquitous and intelligent environment. They
explore ways of presenting conceptual designs as investigations and processes rather than as finite things in
hemselves. Each proposal is a radio, an interface between the electromagnetic environment of hertzian space
nd people. Each explores different forms of realism: technical, functional, social and psychological.
osals are not intended for mass-production or even prototyping, but for mass-consumption through publication
nd exhibition. They ask questions rather than provide answers and should stimulate discussion in the way a film
r novel might. Each focuses on different design issues. THIEF OF AFFECTIONS explores designing role models and
psychosocial narratives. From ELECTROCLIMATES emerges ideas for genotypes, pseudo-interviews, and poetic
products. WHEN OBJECTS DREAM... offers alternative conceptions of the smart object as dreamy object, and new
ools like the gaussmeter for mapping hertzian space. TUNEABLE CITIES explores overlapping electromagnetic and
rban spaces using a car and scanner to experience a city. Architectural models emphasise radio as environment
ather than medium, and video stresses the design of experience rather than that of objects. FARADAY CHAIR
nvestigates a conceptual approach to the aesthetics of hertzian space and the object.

ELECTROCLIMATES: ABSTRACT RADIO

This proposal developed from my desire to create a post-optimal object that answered aesthetic needs within a context of everyday life. It would be an aid for poetically inhabiting the electrosphere, a contemplative object revealing the hertzian nature of our environment.

I began the investigation with the realisation that hertzian space is not isotropic but has its own electromagnetic 'climate' which is related to an electrogeography defined by wavelength, frequency, and field strength, and interacts with urban and natural environments as discussed above in **Electroclimates** in **HERTZIAN SPACE.**

To make visible minute atmospheric changes, antique meteorological devices such as barometers, hygrometers and thermometers, often use unusual means which reveal the sensuous materiality of space. For example, to indicate humidity, some hygrometers exploit the expansion of hair and skin when moisture is absorbed, and the transparent design of early barometers show they measure the 'weight of the sky'. My object would be an electronic relative of the early radio detectors which also employed ingenious means of indicating the presence of radio waves **(FIG. 7.1).**

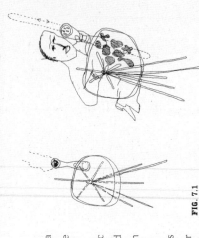

FIG. 7.1
Concept Sketch.

FIG. 7.2
Final design for Electroclimates.

My original proposal was a radio that converted electroclimatic changes into abstract sounds using a wideband radio scanner. It would allow one to notice patterns and become familiar with the flow of activity within a particular area. On another level **ELECTROCLIMATES** is a response to the communications that invade domestic spaces. When a scanner is used in the privacy of the home to listen in on a telephone conversation outside it, the user is seen as the invader but, seen from another viewpoint, the radio signals from cellular telephones are invading the home. **ELECTROCLIMATES** uses an aesthetic language to gently draw attention to this new and problematic interface between private and public space.

An opportunity to use LCD screens arose. The first reaction was hesitancy. Screens are like 'supermatter': once switched on all attention turns to them, and their material qualities are demoted to the status of package or container as the viewer searches for the real content, information. Unlike sound, which can be non-directional, screens tend to give a space a specific orientation.

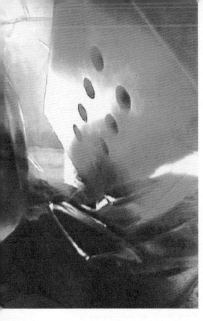

FIG. 7.3
Still from PILLOW TALK video.

FIG. 7.4-5
Stills from PILLOW TALK video.

To explore subtle and evocative uses for the monitor as a material, a way to use a screen to communicate gently and impressionistically, was found by experimenting with different plastics. When thin sheets of sanded fluorescent polycarbonate are held close to the screen they interact with its light to produce a hazy effect. I made a simple animation that slowly changed colour and gently pulsated.

I then explored the physical nature of ELECTROCLIMATES through rough sketches and scale models. By arranging the screen horizontally, it could be viewed from any direction, overcoming the dominance screens have on the layout of rooms. The two main areas of investigation were into ELECTROCLIMATES as a piece of furniture such as a small table or a ceiling fixture, and a portable device.

I decided not to simulate industrial production but make a hand-made genotype, an object designed to communicate the essence of the idea that could later be developed for mass- or batch-production if the occasion arose. The entire object would be made from one material so that the screen appeared to dissolve into it.

As I experimented with the scanner and considered how ELECTROCLIMATES would be used, one time of day became particularly interesting: late at night, as callers sleepily said good night to each other from their beds. This led to the idea of interacting with a product at the moment when boundaries between reality and dreams begin to merge. ELECTROCLIMATES became a 'pillow'.

ELECTROCLIMATES responds to local changes in the radio frequency environment by switching itself on when it detects signals stronger than the general background. It turns electrical space invasions of the home into flickering patterns of light and distorted sounds, when a head is placed on the pillow the distortion clears revealing what is actually being received (for example, telephone conversation, fax transmissions, or garage door openers). Through a slow, gentle interaction the owner would gradually learn to read their electromagnetic environment through the object's responses.

The final design consisted of an LCD screen encapsulated within a fluorescent polycarbonate box which is suspended in a clear PVC inflatable pillow **(FIG. 7.2)** and connected by a lead to a wideband discone aerial. Discone aerials are usually located outside the home; this one is indoors to emphasise that radiowaves are penetrating domestic space.

FIG. 7.6
Still from PILLOW TALK video.

The semi-working object was shown in the MONITOR AS MATERIAL exhibition at the Royal College of Art, which while demonstrating the potential of LCD screens, also provided an opportunity to test public receptivity to the idea of electronic products for answering poetic needs. When fully explained to visitors, ELECTROCLIMATES elicited an enthusiastic response. Without an explanation however, most people saw it only as an exhibition piece rather than a potential product.

As a result of feedback from exhibiting ELECTROCLIMATES, I made a pseudo-documentary video in collaboration with Dan Sellars and Fiona Raby. An elderly woman in her home describes how she thought she would live with an object like ELECTROCLIMATES, how she came by it, when she used it, and what she used it for. We explored where she would keep it, how often she would use it, and how her friends and neighbours might react (**FIGS. 7.3-6**)

The intention was to steer between a number of established approaches: user-testing requires that the object works fully; product clinics test consumer reactions to a product based on how things are now, as are Design Age sessions with the University of the Third Age; 'Informance' aims to persuade an audience that a product fits in and has a place. But here the aim was not to convince an audience of a need, but to draw them into a 'what if...' scenario, a 'value-fiction' to stimulate a desire for change. The interviewee is a knowing participant in a fiction.

In some ways ELECTROCLIMATES 'fails': it is too seductive to be a 'critical design' in that the values it embodies are not strange enough.

WHEN OBJECTS DREAM...

Most people are aware that products like desktop computers, faxes and televisions emit low level electromagnetic (EM) radiation; but it is still unclear if it is harmful. This proposal started as an object for electromagnetic spaces generated by electronic products.

It developed simultaneously along three different paths: ways of sensing and indicating the presence of fields, uses for registering the presence of fields, and the physical nature of the object itself.

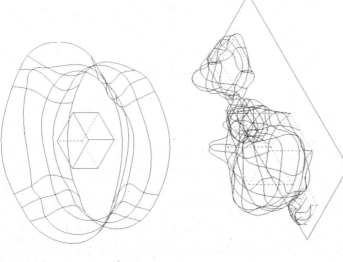

FIG. 7.7
Drawings of fields from a televison and other domestic objects.

My first idea was for small containers that avoided electromagnetic fields. They would be kept on a desktop and move away from the fields they detected. Another idea was to incorporate compasses into a tabletop so that fields from devices placed on it would become visible through the deflecting needles. But compasses are not sensitive enough to be influenced by radiation emitted by computers. Both ideas were for 'enchanted' objects that would mysteriously come alive.

I looked at equipment for measuring VLF (very low frequency) and ELF (extremely low frequency) emissions from products, but these were beyond the budget of this project. So I used a gaussmeter a device for measuring the magnetic component of electromagnetic fields to measure and draw fields produced first by televisions and later by a computer, answerphone, printer and fax machine arranged on a table (**FIG. 7.7**). The gaussmeter revealed an alternative vision of electronic objects as fields, which led to the idea of 'dreamy objects' (see **When Objects Dream...** in **REAL-FICTION** above). It was chosen as the technical basis of the project.

I explored more design ideas: adhesive nipples that vibrated when they sensed fields, warning the wearer to move back, seat backs with vibrating nodules that indicated radiation was passing through the sitter, and parasitical lights that only worked if positioned in fields emitted by domestic products.

Most of these ideas appeared either too whimsical or, in the case of the lights, too feasible. I returned to the idea of the electronic product as a dreamy object and decided to develop an 'object for seeing the dreams of consumer products'. (From this point on, **WHEN OBJECTS DREAM...** was developed simultaneously with **ELECTROCLIMATES** as part of the **MONITOR AS MATERIAL** project.)

A 'glove' was considered first, as though the wearer were caressing the invisible skin of the electronic product, locating its true limit. But this seemed too intimate: there should be more distance between the person and the dreams of products. I decided the device should only work when placed 'at arms length' into the leaky field of an electronic object, a space we can never sense.

The frequency, wavelength and intensity of the victim object's 'dreams' influenced the colour fields and sounds emitted by an LCD screen encased in fluorescent polycarbonate (**FIG. 7.8**). The final object was not made to look like an injection moulding but to appear abstract and brutal. The part of the object touched by

a person was made from a square block of wood, emphasising that the human qualities are not in the form but in what the device does. Two headphone sockets allowed the sounds to be shared.

Like ELECTROCLIMATES, WHEN OBJECTS DREAM... is a semi-working genotype designed for the MONITOR AS MATERIAL exhibition. Its screen shows a video of a computer animation. Ideally it would be presented juxtaposed with a consumer product like a television.

THIEF OF AFFECTIONS

This proposal is based on the realisation, discussed in **Products as Role Models** in **PSYCHOSOCIAL NARRATIVES** above, that electronic products are 'role models' and that when we use them we become the generic user they are modelled on.

THIEF OF AFFECTIONS started with my desire to design an object that embodied an alternative model of a user, a 'perverse' role model. This project is grounded in perversity: not sexual perversion but the desire to rebel, to deny the system the satisfaction of total conformism. Its use would place the user, now a protagonist, into a new relationship with the familiar, providing a new narrative dimension to everyday life.

The approach was similar to those architectural projects described as 'a house for...' where a specific psychological model generates an unusual set of functional requirements. In this case the proposal was to be a walkman for an 'otaku', a term used in Japan to describe an obsessive person, usually male, slightly perverse and socially dysfunctional. The design of the personality became part of the product.

My mis-reading of DOPPLERDANSE(01) (1991) by Steve Mann led to the idea of using radar to caress the internal organs of unsuspecting strangers: the otaku, perversely attempting to experience intimacy by technologically groping the victim's heart **(FIG. 7.9)**, would become a thief of affections. The caress would be converted into vaguely erotic sounds.

The project began to follow two lines of investigation: a technological investigation of the 'caress' and how 'affection' could be stolen, and an exploration of the physical nature of the 'walkman'.

FIG. 7.8
LCD screen encapsulated in flourescent polycarbonate.

At first I mistakenly assumed that different frequencies of radar penetrate the body to different depths rather than reflecting off its surface. I considered ultrasonic scanners, which do penetrate the body to different degrees, but they require the transducer to be placed in contact with a gel spread on the skin. When I discovered that the body gives off a very weak electromagnetic field, the idea of the thief stealing very weak radio emissions from the body appeared feasible. But these signals are so weak that highly specialised and bulky equipment would have to be used, and at very close range. And the technology had to be believable if the proposal was to be a 'value-fiction' not a 'science-fiction'.

I then considered the physical nature of the device. It could easily become a black box, an alternative walkman, or be incorporated into existing objects like clothing or the body (as an electronic tattoo for instance). Although it could be argued that the experience produced by the electronic technology alone is the product, I felt that the nature of the object itself was as important. The juxtaposition of experience and object counted in the sense discussed in **Juxtaposition** in **THE ELECTRONIC AS POST-OPTIMAL OBJECT** above.

To generate the object I looked again at the character of the thief. Being 'perverse', the thief would resist the trend towards miniaturisation and absorption of electronics into existing objects or the body. Hyper-sensitive to technology penetrating his own body, he would favour glasses over contact lenses, baggage and walking sticks over pockets. The final object would be separate from the body, like a walking stick.

The physical configuration of the object was then explored through sketches.

Considering how close technology can come to the body before it becomes invasive led to the pacemaker, the ultimate technological invasion of the body, which transmits weak radio signals. The thief would steal the radiation given off by the artificial heartbeat of a radio heart, <O2> becoming the 'Thief of (Radio) Affections', placing himself in a new form of intimacy with his unsuspecting victims. **(FIG. 7.10)**.

The weak radio signals emitted by the pacemaker could be picked up by a test-probe of the kind normally used to measure stray electromagnetic emissions from domestic appliances. Although relatively expensive they could form a technical basis for the device. Essentially it would be a radio tuned to a very narrow part of the electromagnetic spectrum.

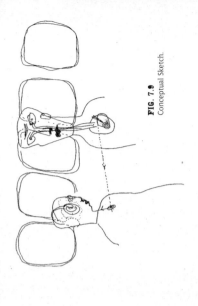

FIG. 7.9
Conceptual Sketch.

FIG. 7.10
Object of desire: pacemaker.

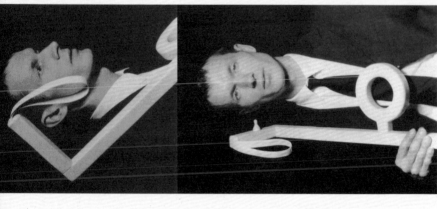

FIG. 7.11-12
Preliminary studies for a portrait of the THIEF OF AFFECTIONS.

I then started exploring the physical design through scale models. I decided it should be audacious. The thief would display the fact he was engaged in some unconventional activity, but simultaneously provide himself with a conspicuously voyeuristic mask. The object became more decadent, like a riding crop. Originally conceived as being made from neutral materials it was now to be made of leather, referring to both its status as luggage and skin. The thief would be conservative, so brown leather was chosen.

While developing the device's physical aspects I was torn between making it a utilitarian tool, a prop for a specific narrative, or an abstract design object. The tool was rejected on the grounds that the contrast with its non-utilitarian function would be two obvious. The first leather version worked well as a film prop, but less well as an object in its own right, the object's slightly antique appearance, meant it could be mistaken for a curious antique object, and its poor workmanship made it look too flimsy to be convincing. A second version was designed to explore the expressive possibilities of highly synthetic materials such as upholstered technical fabrics creating a less familiar image for the device. This version was an abstraction of the earlier version and it was covered in thin foam and flesh coloured fabric suggestive of prosthetic limbs and sex toys. The abstract form restrains the overt imagery of the fabric. It yields few clues about its function, other than it is easily wiped clean. Two straps and a plastic ear-nipple offer more suggestions.

The THIEF OF AFFECTIONS would be presented in a 'shoe box', marked with the size of the object (S, M, L, or XL). And the device itself would be designed to last for as long as possible and would not be adjustable. The weak signals picked up by the device would be converted into sounds played through an 'ear-nipple' of 'prosthetic beige' plastic. The sounds, (developed by Jayne Roderick) would range from vaguely masculine to vaguely feminine depending on preference.

The object would be carried, swinging by the side, and would be hoisted to the shoulder for use. The change in position would cause a tilt switch to activate an automatic scan of a range of pacemaker frequencies, locking onto any close signal. Interaction with the object is minimal. Interaction *through* the object with unsuspecting victims is more important. The device could also be rented for short periods to provide new experiences.

From the start the object was viewed as neither a conventional non-working mock-up nor a fully working prototype, but as a prop for a series of photographs inspired by a mixture of anthropological and medical photography. I worked with photographer Lubna Hammoud developing a series of scenarios which focus on the psychology of the owner as well as the object, emphasising the psycho-social narrative possibilities of an electronic object as a role model **(FIG. 7.11-12)**.

The strangeness of the behavioural model embodied in this proposal draws attention to the fact that all electronic products embody models about behaviour and it questions just how distinct our own identity is from those embodied in the electronic objects we use.

TUNEABLE CITIES

TUNEABLE CITIES investigates overlapping electromagnetic, urban (and natural) environments. It uses the car as a found environment/object, the product designer's entry point into urbanism. With its built-in radio, telephone, navigator and even television, the car is already an interface between hertzian and physical space.

FIG. 7.13
Sketch of local radio blocks.

The proposal began during a drive across Ireland. While the changing landscape is visible, the car isolates its occupants from corresponding cultural changes and the changing radioscape. The car radio reinforces this by automatically re-tuning to a station selected at the start of the journey. But if the radio instead tunes automatically to local stations, changes in the landscape would be matched by changes in culture, interest and dialect revealing the vernacular qualities of hertzian space.[03] The car would link its occupants to the environment rather than isolating them from it **(FIG. 7.13)**.

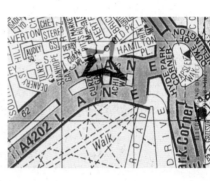

FIG. 7.14
Suspicious click.

This proposal was connected with an earlier unsubmitted idea for a competition to design a monument for Shepherd's Bush Green roundabout in London. My proposal was an abstract radio sculpture which would produce a radio environment designed to penetrate passing cars. Other transmitters could be located around London, so that when the car radio is tuned to a 'public utility' frequency, the car environment becomes a mobile capsule of abstract sound as it moves through the city.

FIG. 7.15
Illegal bugging device.

FIG. 7.16
Video still from RADIO BIRDS video.

The next stage, in collaboration with Fiona Raby, used a car, wideband radio scanner, scanner directory, and street map to search for interesting overlaps between electromagnetic and urban environments in London. The scanner read interstitial urban radio spaces. British Telecom's research laboratories provided computer models of interactions between cellular phone signals and urban environments which encouraged us to think of these radio spaces as environments.

The scanner was programmed to search for illegal bugging devices as we drove around Park Lane, Gray's Inn, Fleet Street, and Mayfair. It registered possible bugs by momentarily 'clicking' on known frequencies which were then marked on a map **(FIG. 7.14)**. We found our first definite bug in Mayfair. By parking close to two edges of a building, we picked up a bugged conversation which we registered on the map **(FIG. 7.15)**.

We then programmed the scanner to receive transmissions from babycoms (baby monitoring intercoms) and drove to suburban Chiswick, a district favoured by new families, and were surprised at the extent babycoms were transmitting domestic sounds into the street. We marked our findings on a map and found that on some streets almost half the houses transmitted domestic soundscapes.

We video-recorded some of the areas we had explored, digitised the tapes and produced short video clips visualising the radio events as environments and experiences. This stage experimented with different languages of representation. The videos aimed to convey a sense that radio is not only a medium, but is environmental: it can be occupied, extending conventional architectural spaces to blur boundaries between private and public.

Alternative sources of radio were explored. Two further possibilities emerged: mobile fields of abstract sound produced by radio-tagged birds, and natural radio produced by atmospheric events. For RADIO BIRDS **(FIG. 7.16)**, birds (possibly already tagged by scientists) would become radios generating seasonal fields of abstract sound as they migrate. They would either transmit directly to the car or reflect signals transmitted to them from arrays of antennas positioned at ground level or on buildings, which would in turn be transmitted to the car. Alternately arrays of antennas, functioning as perches would amplify and transmit the signals to passing cars **(FIG. 7.17)**.

101 HERTZIAN TALES AND SUBLIME GADGETS

PUBLIC UTILITY would consist of zones of speed-trap radar. When cars entered these zones their presence would affect the signal they were receiving so that the sound environment of the cars would be directly affected by the cars using the roundabout.

We then used the language generated by the earlier video experiments to present a development of the PUBLIC UTILITY and RADIO BIRDS. We chose two more London sites: Waterloo for PUBLIC UTILITY and Trafalgar for RADIO BIRDS. Three-dimensional architectural models using synthetic materials and silk-screened maps were also made for each environment to reinforce the shift in emphasis from radio as energy to radio as space.

The final stage of TUNEABLE CITIES shifted attention back to the physical nature of the car radio itself. At first the radio was going to be a clip-on car accessory, an alternative use of the existing car radio slot, or something to do with adhesive patches and tax discs (FIG. 7.18), but I felt that the essence of this project is the re-engineering of a radio as design's potential for subversion lies in the product's function rather than its form.

Ideally the design of the radio would have been given to an established commercial design practice and aimed at a particular market. As that was beyond the scope of this project, images of existing car radios were modified to show alternative functions organised as legal/illegal, urban/rural and private/public pairs (FIG. 7.19). Some presets for sferics, bugs, babycoms, PUBLIC UTILITIES and RADIO BIRDS were also included.

The design process behind this proposal acknowledges the electromagnetic spectrum as a social space, where new definitions of private and public are currently being worked out. Illegal bugging devices, and babycoms which unintentionally act as bugs, provide extensions of private spaces into the public realm. Embassies, legal districts, and suburbs are already part of a tuneable city. RADIO BIRDS explores relationships between people and an artificial nature mediated by cars, while PUBLIC UTILITY draws attention to the disembodied public space shared by transient mobile communities of car drivers.

The TUNEABLE CITIES proposals question what part of a design process needs to be communicated, and how. They take the car as a found environment/object and revisit the city using mass-produced products to explore public and private space, artificial nature, public art, and overlaps between electromagnetic and urban environments. They suggest a role for electronic products as shapers of urban experience.

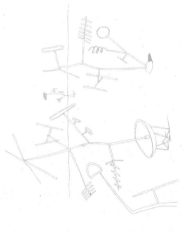

FIG. 7.17
Sketch of RADIO BIRDS antennas.

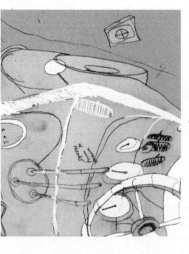

FIG. 7.18
Sketch of car radio.

FIG. 7.1⁹
Treated car radio image.

FARADAY CHAIR: NEGATIVE RADIO

Science tells us that radio is everywhere at all times. Whenever an electron changes its motion, the disturbance brings about an electromagnetic wave with radio frequency. This means that as you turn the pages of Radiotext(e) you're creating radio waves. So handle this anthology carefully – radio waves never die.

N. STRAUSS, RADIOTEXT(E)

During a project about 'electronic space' I realised that today all space is electronic, and that the challenge to designers is to create an 'empty' space, a space that has not existed for most of this century due to the explosion of uses for the electromagnetic spectrum. This proposal is not concerned with dramatic aesthetic expressions of electronic phenomena but with providing a 'conceptual buzz'.(04)

My proposed object for presenting a non-electronic, radio-free volume would use a faraday cage to show the ubiquitous nature of radio space and make perceptible the absence of radio. The object would ask: if the inside is empty, what is outside?

My first ideas were for literal faraday cages which containing something natural like fruit, would imply that natural objects needed to be protected from electromagnetic pollution. But the mesh was easily interpreted as a decorative device and, although the earth lead for the cage conveyed an impression of electrical functionality, it made it look scientific in a quaintly Victorian way. So I sought less technical-looking shield materials and a more powerful subject for enclosure.

The sculptor Jannis Kounellis uses a harsh steel bed frame to represent the absent human. This made me realise that the human body should be what is contained, and that even when unoccupied the object should still refer to the body. In Kounellis' piece the bed's power as an image stems from both its human scale, and the fact that we are born and die on it.

The use of chairs to express prevailing values and ideas about design is well established. Chairs also echo the human body and can communicate new images of man. For example, the bean bag expressed the new informality of the 1960s. A FARADAY CHAIR could provide a new image of the technological person: not of

FIG. 7.20
Final version with occupant.

a cyborg fusing with technology, or of a master of technology, but of vulnerability and uncertainty about the long-term effects of the technologies now so enthusiastically embraced.

While exploring ideas for the FARADAY CHAIR, and as a reaction to the awkwardness of the chairs, the day bed suggested an elegant image of an object whose occupant escaped not into a fantastic world of VR but to enjoy the conceptual buzz of a pure electronic radiation-free space. But the calm repose of the day bed seemed too decadent. The awkwardness of the chair conveyed desperation, and its cramped space lent preciousness to the pure space it contained. The final object even if a daybed, should not be too luxurious as it might appear utopian. There should be something not quite right about it. Its origins should appear uncertain.

By this stage I had moved from metal meshes and sheets to tinted acrylics or glass, and to silkscreened conductive inks or conductive films and coatings which were visually transparent but radio-opaque. I rejected fabrics because radio-opaque versions were only available as visually opaque and it was important to highlight the difference between visual and radio transparency.

Although the final object was a smaller version of a day bed, requiring the occupant to adopt a foetal position (**FIG. 7.20**), I kept the title FARADAY CHAIR to suggest that, once electromagnetic fields are taken into consideration, conventional assumptions about everyday objects need to be re-examined. The final proposal is a compromise between the awkwardness of the chair and the abstract elegance of the day bed. A snorkel mouthpiece attached to a silicone air tube was added to hint at the darker psychological side of the proposal and counter the object's formal elegance (**FIG. 7.21**).

I developed a series of photographic scenarios with Lubna Hammoud for this object too. The emphasis was on portraying the vulnerability of its user. Photography was chosen over video as very specific moments could be constructed and more left to the viewer's imagination.

This proposal draws attention to aesthetic differences between sensual and conceptual approaches to the electronic object. The object is stripped back to its essentials, learning from ELECTROCLIMATES, the FARADAY CHAIR is less seductive and more difficult to accept.

MATERIAL TALES

The design proposals described in this chapter function as conceptual test-pieces that, through their strangeness, make visible some of the social and psychological mechanisms that shape aesthetic experiences of everyday life mediated by electronic products. Their apparent unusability creates a heightened sense of 'distance'. This can be because the objects do not work technically or, preferably, because they are conceptually difficult to assimilate. Through use, or at least by modelling a scenario of use in the mind, the observer discovers new ways of conceptualising reality. They challenge how we think about extensions to our 'selves' in ways that do not simply magnify but, rather, transform our perception and consciousness of our relation to our environment.

They are material tales that allow complex interactions between reality and imagination. Driven by poetry, imagination and intuition rather than reason and logic, they have their own sense, an alternative to our everyday scientific-industrial one. These are tales about the space between rationality and reality, which in an industrial society have come to be synonymous.

FIG. 7.21
Close-up.

THE CONCLUSION

This project set out to develop a design approach for producing conceptual electronic products that encourage complex and meaningful reflection on inhabitation of a ubiquitous, dematerialising and intelligent environment: a form of social research to integrate critical aesthetic experience with everyday life.

Whereas architecture and fine art often refer to popular culture, industrial design *is* popular culture. Its language is accessible and appeals to the senses and imagination rather than the intellect. I hope in my approach I have retained the popular appeal of industrial design while using it to seduce the viewer into the world of ideas rather than objects. Industrial design locates its object in a mental space concerned with identity, desire and fantasy, and shaped by media such as advertising and television. Again, I hope this remains intact, but is subverted to challenge the aesthetic values of both consumers and designers.

One result of this research is a toolbox of concepts and ideas for developing and communicating design proposals that explore fundamental issues about how we live amongst electronic objects. The most important elements of this approach are: going beyond optimisation to explore critical and aesthetic roles for electronic products; using estrangement to open the space between people and electronic products to discussion and criticism; designing alternative functions to draw attention to legal, cultural and social rules; exploiting the unique narrative possibilities offered by electronic products; raising awareness of the electromagnetic qualities of our environment; and developing forms of engagement that avoid being didactic and utopian. The emphasis on behaviour as narrative experience stimulated by the design of new functions links these proposals more to the complicated pleasure of literature and film than sculpture — the traditional reference for industrial design.

But perhaps the most important conclusion is that design is existential and cannot ignore its complicated relationship to people and their mental lives. This project offers one possible approach to allow some of the more complicated and difficult aspects of our relationship to electronic objects to be reflected in future designs.

BIBLIOGRAPHY

J. ABALOS & J. HERREROS, "TOYO ITO: LIGHT TIME" IN
EL CROQUIS 71, (1995).

H. ALDERSEY-WILLIAMS, THE NEW CRANBROOK DESIGN
DISCOURSE. NEW YORK: RIZZOLI (1990).

M. AMIS, EINSTEIN'S MONSTERS.
LONDON: PENGUIN (1987).

A. APPADURAI (ED), THE SOCIAL LIFE OF THINGS:
COMMODITIES IN CULTURAL PERSPECTIVE.
CAMBRIDGE: CAMBRIDGE U.P (1986).
— "INTRODUCTION: COMMODITIES AND THE POLITICS OF VALUE"
IN APPADURAI.

APPLE COMPUTER, KNOWLEDGE NAVIGATOR.
CHI 92 SPECIAL VIDEO PROGRAM ISSUE 79 (1992). SIGGRAPH
VIDEO REVIEW (VIDEO), 1987.

D. AUGAITIS & D. LANDER (EDS.), RADIO RETHINK.
BANFF: WALTER PHILIPS GALLERY (1994).

G. BACHELARD, THE POETICS OF SPACE.
BOSTON: BEACON (1969).

N. BAKER, THE MEZZANINE. CAMBRIDGE: GRANTA (1986).
ROOM TEMPERATURE. LONDON: GRANTA (1990).
VOX. LONDON: GRANTA (1992).

J. G. BALLARD, CRASH. LONDON: PALADIN (1990).

R. BANHAM, DESIGN BY CHOICE. LONDON: ACADEMY (1981).
THEORY AND DESIGN IN THE FIRST
MACHINE AGE. LONDON, ARCHITECTURAL PRESS, 1960

R. BARTHES, MYTHOLOGIES. LONDON: PALADIN (1989).

JEAN BAUDRILLARD, FOR A CRITIQUE OF THE
POLITICAL ECONOMY OF THE SIGN.
ST. LOUIS: TELOS (1981).
"THE ECSTASY OF COMMUNICATION" IN H. FOSTER, THE
ANTI-AESTHETIC. SEATTLE: BAY PRESS (1983).
THE ECSTASY OF COMMUNICATION.
NEW YORK: SEMIOTEXT(E) (1988).
"THE SYSTEM OF OBJECTS" IN THACKARA (1988).
THE REVENGE OF THE CRYSTAL.
LONDON: PLUTO (1990).
THE TRANSPARENCY OF EVIL.
LONDON: VERSO (1993).
"XEROX AND INFINITY" IN BAUDRILLARD (1993).
"THE METAFUNCTIONAL AND DYSFUNCTIONAL SYSTEM:
GADGETS AND ROBOTS" IN BAUDRILLARD (1996).
THE SYSTEM OF OBJECTS.
LONDON: VERSO (1996).

C. BECKER (ED), THE SUBVERSIVE IMAGINATION.
NEW YORK AND LONDON: ROUTLEDGE (1994).
"HERBERT MARCUSE AND THE SUBVERSIVE POTENTIAL
OF ART" IN BECKER (ED).

C. BEEVOR, "BETWEEN HERE AND NOW" IN NATO.
LONDON: ARCHITECTURAL ASSOCIATION (1983).

G.BENDER, & T.DRUCKREY (EDS.), CULTURE ON THE BRINK.
SEATTLE: BAY PRESS, (1994)

M. BERTOMEN, TRANSMISSION TOWERS ON THE L.I.E.
(LONG ISLAND EXPRESSWAY). NEW YORK:
PRINCETON ARCHITECTURAL PRESS (1991).

A. BIRNBAUM, "FEXY FACTS" IN MEDIAMATIC 5 (4) (1991).

D. J. BOLTER, TURING'S MAN. LONDON: PENGUIN (1993).

N. BOLZ, "THE MEANING OF SURFACE" (1992) IN NETHERLANDS
DESIGN INSTITUTE (1994).

O. BOUMAN, REAL SPACE IN QUICK TIMES:
ARCHITECTURE AND DIGITISATION.
ROTTERDAM: NAI (1996).

A. BRANZI, THE HOT HOUSE. LONDON:
THAMES AND HUDSON (1984).
LUOGHI ANDREA BRANZI THE COMPLETE
WORKS. LONDON: THAMES AND HUDSON (1992).
"CRISIS AND GROWTH: INDUSTRIAL DESIGN AND THE
CREATIVE TRADITION" IN MARZANO (1995).

R. BRIDGMAN, ELECTRONICS. LONDON:
DORLING KINDERSLEY (1993).

C. BURNS ET AL, "ACTORS, HAIRDOS & VIDEOTAPE – INFORMANCE DESIGN"
IN HUMAN FACTORS IN COMPUTING SYSTEMS
CHI '94 CONFERENCE PROCEEDINGS.
NEW YORK: ACM PRESS (1994).

H. CALAS & N. CALAS, TAKIS. PARIS: ÉDITIONS GALILÉE (1984).

G. CEPPI, "PLAYING WITH TECHNOLOGY" IN MODO 136 (1991).

C. CHANT (ED), SCIENCE, TECHNOLOGY AND EVERYDAY
LIFE 1870—1950.
LONDON: ROUTLEDGE (1989).

G. CHAPMAN, "MAKING SENSE OUT OF NONSENSE" IN BENDER, &
DRUCKREY (1994).

W. CHOE, TOWARD AN AESTHETIC CRITICISM OF
TECHNOLOGY. NEW YORK: PETER LANG (1989).

K. CLARKE, POSITIONING IN RADIOGRAPHY. LONDON:
HEINEMANN (1967).

R. COLEMAN, DESIGN RESEARCH FOR OUR FUTURE
SELVES (2). LONDON: ROYAL COLLEGE OF ART (1993/4).

M. CSIKSZENTMIHALYI & E. ROCHBERG-HALTON, THE MEANING OF
THINGS. CAMBRIDGE: CAMBRIDGE U.P (1981).

N. DALY & B. LYONS, "THE LOST CIVILIZATION OF LLHUROS:
THE FIRST MULTIMEDIA EXHIBITION IN THE GENRE
OF ARCHAEOLOGICAL FICTION" IN LEONARDO 24 (3) (1991).

M. DELANDA, WAR IN THE AGE OF INTELLIGENT
MACHINES. NEW YORK: ZONE (1991).

M. DE LUCCHI, & S. MARZANO, WORKSHOP.
EINDHOVEN: PHILLIPS/OLIVETTI (1994).

J. DEWEY, ART AS EXPERIENCE. NEW YORK:
MINTON, BALCH (1958).

C. DIKE, CANE CURIOSA: FROM GUN TO GADGET.
PARIS: ÉDITIONS DE L'AMATEUR (1983).

D. DOLLENS, "TOWARD NINJA ARCHITECTURE/HYPERARCHITEXTURE" IN
OFF RAMP 1 (4) (1991).

P. DORMER, THE MEANINGS OF MODERN DESIGN. LONDON:
THAMES & HUDSON (1990).

A. DUNNE & F. RABY, "FIELDS AND THRESHOLDS" IN ARCHITECTS IN
CYBERSPACE. ARCHITECTURAL DESIGN PROFILE NO. 118.
LONDON: VCH (1995).

A. DUNNE & W. GAVER, "THE PILLOW: ARTIST-DESIGNERS IN THE DIGITAL
AGE". IN HUMAN FACTORS IN COMPUTING
SYSTEMS CHI '96 CONFERENCE
PROCEEDINGS.
READING, MA: ADDISON WESLEY (1996).

J. K. FEIBLEMAN, TECHNOLOGY AND REALITY.
THE HAGUE AND LONDON: NIJHOFF (1982).

A. FORTY, OBJECTS OF DESIRE. LONDON:
THAMES AND HUDSON (1986).

M. FOUCAULT, THE ORDER OF THINGS.
LONDON AND NEW YORK:
TAVISTOCK/ROUTLEDGE (1970).

K. FRAMPTON & S. KOLBOWSKI (EDS.), IDEA AS MODEL.
NEW YORK: RIZZOLI (1981).

J. FRAZER, THEMES VII: AN EVOLUTIONARY
ARCHITECTURE LONDON:
ARCHITECTURAL ASSOCIATION (1995).

S. FREUD, JOKES AND THEIR RELATION TO THE
UNCONSCIOUS. LONDON: ROUTLEDGE (1966).

M. FUJIHATA, FORBIDDEN FRUITS. TOKYO: LIBRO PORT (1991).

P. GARNER, PHILIP GARNER'S BETTER LIVING
CATALOGUE. LONDON: SIDGWICK & JACKSON (1983).
UTOPIA OR BUST. LONDON: PLEXUS (1985).

R. GEUSS, THE IDEA OF A CRITICAL THEORY.
CAMBRIDGE: CAMBRIDGE U.P (1981).

V. GRASSMUCK, "OTAKU" IN MEDIAMATIC 5 (4) (1991).

P. GRIFFITHS, MODERN MUSIC.
LONDON: THAMES & HUDSON (1986).

T. HAWKES, STRUCTURALISM AND SEMIOTICS.
LONDON: ROUTLEDGE (1992).

M. L. HEIVLY & M. REED, "THE SPACE BETWEEN THE REAL AND THE
IMAGINED: MICROWAVE SCULPTURE IN DEEP SPACE" IN
LEONARDO 25 (1) (1992).

C. HUBERT, "THE RUINS OF REPRESENTATION"
IN FRAMPTON & KOLBOWSKI (1981).

K. IRIE, (1988) "COMPUTER CRASH BY DESIGN" IN TELESCOPE,
MAY/JUNE.
"AA INTERMEDIATE UNIT 3: OBJECTS IN THE LANDSCAPE" IN
AA FILES 25 (1993).

J. ITO, "ARCHITECTURE IN A SIMULATED CITY" IN EL CROQUIS 71 (1995).

P. JOHNSON, MARXIST AESTHETICS: THE FOUNDATIONS WITHIN EVERYDAY LIFE FOR AN EMANCIPATED CONSCIOUSNESS. LONDON: ROUTLEDGE & KEGAN PAUL (1984).

K. KAWAKAMI, 101 UNUSELESS JAPANESE INVENTIONS. LONDON: HARPER COLLINS (1995).

S. KING, "SCANNER" IN MEAD (NO DATE).

J. KIPNIS, "ATM COMPETITION" IN OFF RAMP 1 (4) (1991).

B. KLÜVER, "ARTISTS, ENGINEERS, AND COLLABORATION" IN BENDER & DRUCKREY (1994).

K. KRIPPENDORFF & R. BUTTER (EDS.), DESIGN ISSUES 5 (2) (1989).

KUNSTFLUG, "DESIGN WITHOUT AN OBJECT" IN OTTAGONO 100, (1991).

L. KURGAN, "YOU ARE HERE: INFORMATION DRIFT" IN ASSEMBLAGE 25. (1995).

D. LANDER, "RADIO AND BEANS" IN AUGAITIS & LANDER (1994).

S. MANN, "DOPPLERDANSE: SOME NOVEL APPLICATIONS OF RADAR" IN LEONARDO 25 (1) (1992).

E. MANZINI, THE MATERIAL OF INVENTION. MILAN: ARCADIA (1986).

—   "PROMETHEUS OF THE EVERYDAY: THE ECOLOGY OF THE ARTIFICIAL AND THE DESIGNER'S RESPONSIBILITY" IN DESIGN ISSUES 9 (1) (1992).

—   "DESIGN, ENVIRONMENT AND SOCIAL QUALITY: FROM 'EXISTENZMINIMUM' TO 'QUALITY MAXIMUM'" IN DESIGN ISSUES 10 (1) (1994).

—   UNPUBLISHED PAPER "DESIGN RESEARCH FOR A SUSTAINABLE DEVELOPMENT" DELIVERED AT DESIGN/RESEARCH CONFERENCE, ROYAL COLLEGE OF ART, MAY 1994 (1994).

E. MANZINI & M. SUSANI (EDS.), THE SOLID SIDE. NETHERLANDS: V+K (1995).

V. MARGOLIN & R. BUCHANAN (EDS.), THE IDEA OF DESIGN. CAMBRIDGE, MA: MIT PRESS (1995).

S. MARZANO, "FLYING OVER LAS VEGAS" IN EUROPEAN DESIGN CENTRE (EDC) NEWS SEPTEMBER (1993).

—   TELEVISION AT THE CROSSROADS. LONDON: ACADEMY EDITIONS (1995).

M. MCLUHAN, COUNTERBLAST. LONDON: RAPP & WHITING (1970).

H. MEAD (ED), TRANCE EUROPE EXPRESS 2. LONDON: BMG (UK) LTD (NO DATE).

H. G. MEAD, MIND, SELF, AND SOCIETY FROM THE STANDPOINT OF A SOCIAL BEHAVIOURIST. CHICAGO: UNIVERSITY OF CHICAGO (1934).

M. MIDEKE, A WHISTLER HUNTER'S GUIDE. PUBLISHED BY THE AUTHOR (1991).

D. MILLER, MATERIAL CULTURE AND MASS CONSUMPTION. OXFORD: BLACKWELL (1987).

L. MILNE & M. MILNE, THE SENSES OF ANIMALS AND MEN. LONDON: ANDRE DEUTSCH (1963).

A. MOLES, "DESIGN IMMATERIALITY: WHAT OF IT IN A POST INDUSTRIAL SOCIETY?" IN MARGOLIN & BUCHANAN (1995).

J. R. MOORE, "EVERYDAY LIFE AND THE DYNAMICS OF TECHNOLOGICAL CHANGE" IN CHANT (1989).

NETHERLANDS DESIGN INSTITUTE, DOORS OF PERCEPTION 1 DOPROM. AMSTERDAM: NDI (CD-ROM) (1994).

B. NICHOLSON, APPLIANCE HOUSE. CAMBRIDGE, MA: CHICAGO INSTITUTE FOR ARCHITECTURE AND URBANISM (1990).

D. NORMAN, THE PSYCHOLOGY OF EVERYDAY THINGS. NEW YORK: BASIC BOOKS (1988).

—   THINGS THAT MAKE US SMART: DEFENDING HUMAN ATTRIBUTES IN THE AGE OF THE MACHINE. READING, MA: ADDISON-WESLEY (1993).

PHILIPS CORPORATE DESIGN, VISION OF THE FUTURE. BUSSUM, NETHERLANDS: V+K (1996).

V. J. PHILIPS, EARLY RADIO WAVE DETECTORS. LONDON: INSTITUTION OF ELECTRICAL ENGINEERS (1980).

D. PORUSH, THE SOFT MACHINE. NEW YORK: METHUEN (1985).

F. POPPER, ORIGINS AND DEVELOPMENT OF KINETIC ART. LONDON: STUDIO VISTA (1968).

—   ART OF THE ELECTRONIC AGE. LONDON: THAMES & HUDSON (1993).

B. RADICE, MEMPHIS. LONDON: THAMES & HUDSON (1985).

A. RICHARDSON, "THE DEATH OF THE DESIGNER" IN DESIGN ISSUES 9 (2) (1993).

R. ROBINSON, BOOK REVIEW IN DESIGN ISSUES 10 (1) (1994).

E. ROCHBERG-HALTON, "THE MEANING OF PERSONAL ART OBJECTS" IN J. ZUZANEK (ED.), SOCIAL RESEARCH AND CULTURAL POLICY. WATERLOO, ONTARIO: OTIUM PUBLICATIONS (1979).

—   CULTURAL SIGNS AND URBAN ADAPTATION: THE MEANING OF CHERISHED HOUSEHOLD POSSESSIONS. UNPUBLISHED PHD DISSERTATION. UNIVERSITY OF CHICAGO (1979).

A. ROSS, "THE NEW SMARTNESS" IN BENDER & DRUCKREY (1994).

M. R. RUBINSTEIN, "PHILIPPE RAMETTE" IN ART FORUM, MAY 1993.

M. SCHAFER, "RADICAL RADIO" IN STRAUSS (1993).

K. SHINOHARA, KAZUO SHINOHARA. NEW YORK: RIZZOLI (1982).

R. SILVERSTONE & E. HIRSCH, CONSUMING TECHNOLOGIES. LONDON: ROUTLEDGE (1992).

C. W. SMITH & S. BEST, ELECTROMAGNETIC MAN. LONDON: DENT (1989).

M. SORKIN, VARIATIONS ON A THEME PARK. NEW YORK: THE NOONDAY PRESS (1992).

—   "ARCHITECTURE RISING LEBBEUS WOODS' AERIAL PARIS PROJECT" IN DIDALOS, DECEMBER 1990.

P. SPARKE, ETTORE SOTTSASS JNR. LONDON: DESIGN COUNCIL (1982).

SPECKLEY PITTMAN PR, MOBILE MOMENTS. LONDON: TELECOM SECURICOR CELLULAR RADIO LTD (1994).

W. STRAUSS (ED.), RADIOTEXT(E). NEW YORK: SEMIOTEXT(E) (1993).

—   "HYDROGEN JUKEBOX" IN STRAUSS.

—   "INTRODUCTION" IN STRAUSS.

J. STURROCK, STRUCTURALISM. LONDON: PALADIN (1986).

SURVEILLANCE TECHNOLOGY GROUP, COVERT AUDIO INTERCEPT. PORT CHESTER: STG (COMPANY BROCHURE).

M. SUSANI, "THE FOURTH KINGDOM" IN OTTAGONO 105. (1992).

—   THE SENSIBLE HOME. LECTURE, DOORS OF PERCEPTION 1 CONFERENCE, AMSTERDAM (1994).

J. TANIZAKI, IN PRAISE OF SHADOWS. LONDON: JONATHAN CAPE (1991).

J. THACKARA, "THE CACOPHONOUS SOUNDS OF TECHNOLOGY" IN AA FILES 11 (1986).

—   NEW BRITISH DESIGN. LONDON: THAMES & HUDSON (1986).

—   DESIGN AFTER MODERNISM. LONDON: THAMES & HUDSON (1988).

B. TSCHUMI, "THE DISCOURSE OF EVENTS" IN THEMES 3. LONDON: ARCHITECTURAL ASSOCIATION (1983).

—   ARCHITECTURE AND DISJUNCTION. CAMBRIDGE, MA: MIT PRESS (1994).

—   EVENT-CITIES. CAMBRIDGE, MA: MIT PRESS (1994).

D. VAN WEELDEN, "MACHINE VOICES" IN MEDIAMATIC 6 (4) (1992).

A. VIDLER, THE ARCHITECTURAL UNCANNY. CAMBRIDGE: MIT PRESS (1992).

P. VIRILIO, THE LOST DIMENSION. NEW YORK: SEMIOTEXT(E) (1991).

—   THE ART OF THE MOTOR. MINNEAPOLIS: MINNESOTA U.P (1995).

B. WAITES, "EVERYDAY LIFE AND THE DYNAMICS OF TECHNOLOGICAL CHANGE" IN CHANT (1989).

P. WEIBEL, "INTELLIGENT AMBIENCE" IN PROMOTIONAL LEAFLET FOR
    ARS ELECTRONICA. LINZ: ORF (1994).

P. WELLER, W. MACKAY & R. GOLD (EDS.), "COMPUTER-AUGMENTED
    ENVIRONMENTS: BACK TO THE REAL WORLD". **IN
    COMMUNICATIONS OF THE ACM** 36 (7) (1993).

K. WODICZKO, **INSTRUMENTS, PROJECTIONS, VEHICLES.**
    BARCELONA: FUNDACIÓ ANTONI TÀPIES (1992).

L. WOODS, **ORIGINS**. LONDON:
    ARCHITECTURAL ASSOCIATION (1985).

P. WRIGHT, "THE POLISCAR: NOT A TANK BUT A WAR MACHINE FOR
    PEOPLE WITHOUT APARTMENTS" IN WODICZKO (1992).

ZIMMERMAN ET AL, "APPLYING ELECTRIC FIELD SENSING TO HUMAN
    COMPUTER INTERFACES" IN **HUMAN FACTORS
    IN COMPUTING SYSTEMS CHI '95
    CONFERENCE PROCEEDINGS.**
    READING, MA: ADDISON-WESLEY. (1995).

ILLUSTRATIONS

DUMB TYPE, PERFORMANCE-PH (1990).
COPYRIGHT DUMB TYPE.

ARCHITEKTUR BÜRO BOLLES • WILSON,
NINJA HOUSE — A COMFORTABLE HOUSE
IN THE ELECTRONIC METROPOLIS (1988).
FIRST PRIZE SHINKENSHIKU COMPETITION.

BEN NICHOLSON, APPLIANCE HOUSE (1990).

TOYO ITO & ASSOCIATES, ARCHITECTS, DWELLING
FOR TOKYO NOMAD WOMAN (1985).
PHOTOGRAPH: TOMIO OHASHI.

TRUTH PHONE", MANUFACTURED BY CCS
INTERNATIONAL LTD.

REAL-FICTION

ANTHONY DUNNE, MONITOR AS MATERIAL (1996).

GREGORY GREEN, NUCLEAR DEVICE #2
15 KILOTONS, PLUTONIUM 239 (1995).

ANDREA BRANZI, GINGER (1980).

MICHELE DE LUCCHI, APPLIANCES (1979).

PHILIPS DESIGN & OLIVETTI,
COMMUNICATOR (FROM WORKSHOP, 1994).

PHILIPS DESIGN & OLIVETTI, GROUP TOOL
(FROM WORKSHOP, 1994).

CINDY SHERMAN, UNTITLED FILM STILL (1978).

PHILIP GARNER, SKEET BOWLING (1985).

PHILIP RAMETTE, OBJECT FOR SEEING ONESELF
WATCHING (1990).

REBECCA HORN, LA FERDINANDA: SONATA
FOR A MEDICI VILLA (1981). COPYRIGHT DACS 1999.

HERTZIAN SPACE

ELECTROMAGNETIC SPECTRUM SHOWING THE
RADIO FREQUENCY SPECTRUM.

KEN KRAYER, TELEPHONE (1988).

STEPHEN J. BIRKILL, SATELLITE FOOTPRINT MAP (1992).
COPYRIGHT STEPHEN J. BIRKILL, ELECTRONICS DESIGN
HOUSE REAL-WORLD TECHNOLOGY LTD.

RADIO PROPAGATION MAP,
COURTESY BRITISH TELECOM RESEARCH LABORATORIES.

X-RAY POSE.

LEFEUVRE'S 'PHYSIOLOGICAL' RECEIVER.

TOYO ITO & ASSOCIATES, ARCHITECTS,
TOWER OF WINDS (1986).

TOYO ITO & ASSOCIATES, ARCHITECTS, DREAMS ROOM
(1991-92). PHOTOGRAPH: NAOYA HATAKEYAMA.

KAZUO SHINOHARA, HOUSE UNDER HIGH
VOLTAGE LINES (1981).

HERZOG & DE MEURON, SIGNAL BOX #4,
AUF DEM WOLF (1991-94).

LOCKHEED F-117 STEALTH FIGHTER.
COURTESY SALAMANDER PICTURE LIBRARY.

LOOP & FRAME ANTENNA.
COURTESY BERNARD BABANI (PUBLISHING) LTD.

DIELECTRIC CLAD ANTENNA.
COURTESY BERNARD BABANI (PUBLISHING) LTD.

GEORGES DROZ-GEORGET, BOBBIN CANE.
COURTESY PATRICK D. GUTKNECHT.

LOW RESISTANCE EARTH SYSTEM (ANTENNA) ARRANGED
WITHIN CONFINES OF SMALL URBAN GARDEN.

BAT BAND CONVERTER.
COURTESY WIMBORNE PUBLISHING LTD.

ELECTROCLIMATES

CONCEPT SKETCH.

FINAL DESIGN.

STILL FROM PILLOW TALK VIDEO.

STILLS FROM PILLOW TALK VIDEO.

STILL FROM PILLOW TALK VIDEO.

WHEN OBJECTS DREAM...

DRAWING OF FIELDS FROM A TELEVISION AND
OTHER DOMESTIC OBJECTS.

LCD SCREEN ENCAPSULATED IN FLOURESCENT
POLYCARBONATE.

THIEF OF AFFECTIONS

CONCEPT SKETCH.

OBJECT OF DESIRE: PACEMAKER.
PHOTOGRAPH: LUBNA HAMMOUD.

PRELIMINARY STUDIES FOR A PORTRAIT OF THE THIEF OF
AFFECTIONS. PHOTOGRAPH: LUBNA HAMMOUD.

TUNEABLE CITIES

CONCEPT SKETCH OF RADIO BLOCKS.

SUSPICIOUS CLICK.

ILLEGAL BUGGING DEVICE.

STILL FROM RADIO BIRDS VIDEO.

SKETCH OF RADIO BIRDS ANTENNAE.

SKETCH OF CAR RADIO.

TREATED CAR RADIO IMAGE.

FARADAY CHAIR

FINAL VERSION WITH OCCUPANT.
PHOTOGRAPH: LUBNA HAMMOUD.

CLOSE-UP.
PHOTOGRAPH: LUBNA HAMMOUD.

IN**D**EX

116